THINK LIKE AN ECOSYSTEM

An Introduction to Permaculture, Water Systems, Soil Science, and Landscape Design

AMÉLIE DES PLANTES

ECOLOGICAL

COPYRIGHT

Acknowledgment

My warmest gratitude for my two children and all the other incredible and inspiring children. Thank you for amplifying my passion and dedication to teaching permaculture, and escalating the need for creating resilient and regenerative ecosystems for future generations.

A Free Gift For You!

FRUIT TREE PRUNING AND SHAPING GUIDE
5 SIMPLE STEPS TO PRUNE FRUIT TREES FOR HUGE HARVESTS AND EASY PICKING

AMÉLIE DES PLANTES

IN THE FRUIT TREE PRUNING AND SHAPING GUIDE, YOU'LL LEARN...

- Why it is important to prune your fruit trees
- 5 Simple steps to prune your fruit trees for huge harvests
- How to choose the best shape for your fruit tree
- How to revive an abandoned or neglected fruit tree
- And so much more...

To receive your Fruit Tree Guide scan the QR code or click the following link

http://ecologicalfoodforest.com/fruit-tree-guide/

Introduction

Lots of things baffle me about humanity, but the biggest is how wasteful we are. People super-size everything without being able to finish it. We buy new TVs, not because the old one was broken but because we want a bigger one. Consumerism is eating the planet.

It's estimated that 30 to 60 percent of our household water is used to irrigate outdoor areas, and 50 percent of that is wasted in runoff (WaterSense, n.d.). In the US, between 125 and 160 billion pounds of food is wasted annually (FoodPrint, 2021). Imagine what we could achieve if we started using these resources.

> *Though the problems of the world are increasingly complex, the solutions remain embarrassingly simple."*
>
> *Bill Mollison*

I was most frustrated at university. Coming from a farming background, I was used to making the most of everything the land could give us. People around me were incredibly wasteful and I didn't want to fall into the same habits. One of my professors was like-minded, and it was thanks to her that I started to pay more attention to renewable, sustainable ecosystems. Her enthusiasm in my determination helped me to understand the concept of making the most of what we have, ironically, something I had had a taste of throughout my childhood, but perhaps at that age, I got the feeling that it could be taken a few steps further.

My professor's wise words were based on a practice from the 70s. She explained that we are so focused on creating beds, irrigation systems, and pest-control methods that we forget how nature does this without intervention. Before 650–600 BC, there was no money and people grew and exchanged goods. We aren't about to go back to such a system, but like countless generations, tribes, and cultures before us, we can create a self-sustaining environment with what nature provides us. She was talking about permaculture.

On her 3,000 square feet of land, my professor was letting nature do what nature does best: using natural water reservoirs, regenerating soil fertility, and allowing animals to control pests as it should be. What's more, she was able to enjoy freshly grown food.

Permaculture is not the same as organic gardening. Both systems are used to grow food, but permaculture goes beyond organic gardening. It involves design and greater responsibility for the waste produced. By mimicking nature, permaculture aims to take advantage of natural resources to provide a sustainable habitat for humans and animals. It sounded simple enough, so during my summers at home, I started to implement the principles of permaculture on a small patch of land on the farm.

When I first began researching permaculture, I was overwhelmed by how complex some people made it. I saw chemical formulas for soil that were enough to put many people off. There was a lot of confusion about what permaculture actually was. Like myself, there seemed to be some confusion between permaculture, organic gardening, and other traditional farming practices. I started peeling back the information, leaf by leaf so to speak, in order to begin my new passion the right way.

Not everything went the way I had initially hoped. There were moments of confusion, even frustration, but the little breakthroughs motivated me to keep going. It was what looked like failures that taught me more about the improvements I could make. Don't worry, I will confess some of my less successful experiences so that you know not to make the same errors!

Five years later, our sustainable ecosystem is providing more benefits than we could have ever imagined. Not only have I learned about things like closed-loop systems and zoning, but now my children are learning the same techniques. We feel healthier, happier, more purposeful, and instead of complaining about the direction the world is going, we feel that our simple solution is making a positive difference.

After founding Ecological Food Forest in 2020, I was thrilled to discover how many like-minded people needed help and advice, whether they had already begun or had no idea where to start. It was clear: Now is the perfect time to share my knowledge and passion so that you can enjoy the same pleasure that my family gets out of permaculture.

We will look at all the principles of permaculture, from observing your environment to planning the different zones. We will learn how to use rainwater, take care of your soil, plant the right plants, and integrate animals. And we are going to do this without the unnecessary jargon that may have previously put you off. Everything we practice will take into consideration the ethical, social, and communal aspects of permaculture.

You don't need to have a large amount of land to practice permaculture. Many people I have helped over the years have just small to average-sized gardens. You don't need fancy tools and expensive equipment. And you don't need to be a landscape designer or even have a green thumb. Your best tools are your enthusiasm and proper explanations.

The first step to creating your sustainable ecosystem is to understand precisely what permaculture is and why this ethical practice differs from organic gardening. There is no need to roll your sleeves up just yet, so sit back and enjoy the first steps to sustainable living with a little bit of a history lesson with a spade full of philosophy.

CHAPTER 1

PERMACULTURE IN A NUTSHELL

TO FULLY UNDERSTAND any new concept, it's always best to start from the very beginning. Even if you have dabbled in the subject and are looking for more advanced advice, knowing the origins of permaculture can help to fully appreciate the concept. It will also help to clear up any misconceptions of our subject.

THE PERMACULTURE BOOTCAMP

There are numerous ways you may have come across the term permaculture. You may have seen the term while researching the internet and thought, what is that? Or, you may have heard it mentioned on popular gardening shows or podcasts. Or, you know someone, a friend maybe, who is raving about their experience and how much it's changed their perspective on life. Somehow, you heard about permaculture. One thing we all have in common is that as soon as you begin practicing permaculture, you can call yourself a permaculturist or, more commonly, a permie. Whether you are starting a new hobby; hoping to optimize your time, space, and resources; or looking for a way to

leave the Earth a prolific and productive place for future generations, all of this can be found through practicing permaculture.

The word permaculture is the combination of permanent and
agriculture. As soon as we make something permanent, we can
see it as sustainable. To get the most accurate idea of what this
means, we will start with the source and how it began.

Since permaculture first appeared in 1978, there have been over
fifty definitions. It's no wonder there is confusion when it gets a
new description at least once a year. The term was coined by Bill
Mollison. He described it as "The conscious design and maintenance of agriculturally productive systems which have the diversity, stability, and resilience of natural ecosystems." He saw it as a
way of people and landscapes integrating for food, energy, and
shelter.

The inspiration for permaculture came nearly two decades
earlier. The Australian was watching marsupials in the
Tasmanian rainforests. He marveled at how the animals and the
ecosystem were interconnected and believed people could
recreate the same thing.

While Mollison gave permaculture its name and turned it into
today's modern movement, the practices and ethics have been
around for centuries. When the Spanish first arrived in the
Americas, they noticed how the indigenous people lived with the
plants and animals and had grown together into one ecosystem.
They used land management to maximize the growth of plants
for food, medicine, and tools. Over the centuries, science has
taken over and we have lost a lot of the balance that the indigenous people enjoyed. Through permaculture, Mollison was able
to bring the practices of the indigenous people back to the fore-
front of gardening practices.

Mollison met David Holmgren while teaching biogeography and environmental psychology at the University of Tasmania. Together, they established the practices that would become the global standards of permaculture. In 1979, Mollison helped found the first Permaculture Institute, with the first students graduating in 1981. By 2001, over 300,000 students had graduated worldwide.

THE BASIC CONCEPTS OF PERMACULTURE

Remember, we are just scratching at the topsoil of permaculture, and throughout the chapters we will go much further into the bedrock of these concepts. For now, let's take a quick look at the basic concepts that permies soon start to live by.

Closed-loop systems

A closed-looped system is one that is sustainable because it provides its own energy. For example, you would grow your own crops for animals or feed them with leftover kitchen scraps. In turn, you would use animal manure as a fertilizer, and so the loop is closed. A permie's motto is to turn waste into resources. Solar energy is free and one of the biggest wasted resources we have. Instead, people are still using energy sources that cause pollution. As we saw in the introduction, water and food waste are also causing indescribable damage to the planet, and it doesn't have to be this way.

Perennial crops

Let's begin with the permie's drawback with annuals. Annual plants and flowers have a survival strategy of producing a high number of seeds because they will die within a year. They grow incredibly quickly and this requires a lot of nutrients from the soil. Roots of annuals don't grow deep enough to absorb the nutrients further down in the soil, so tilling or preparing soil annually or biannually is necessary, but this isn't good for the soil. Perennials are slow-growing with roots that are often longer than the plants themselves. Deep roots require less fertilizer and water. Not only can they take advantage of the nutrients further down, but they also don't require the same level of nutrients as fast-growing annuals. What's more, perennials grow during the summer, growth slows down in winter and commences again when the weather starts to warm up—they are permanent.

Multiple functions

Every part of your sustainable ecosystem will eventually have more than one function. A chicken coop can be solar-powered and have a roof designed to harvest rainwater. The rainwater you collect can be used for the chickens, irrigation, aquatic food plants, even edible fish such as trout and silver perch. Because of this "stacking function," it's essential that your environment is designed in such a way that allows multiple functions to be incorporated.

Eco-earthworks

On the note of the design, eco-earthworks refers to the design of a landscape that allows for the use of every drop of rainwater. If you are like me, you might have started your permaculture design by plotting where you want your herb garden, where the vegetables will be, and where your coops and pens will go. Sadly, you might have to ditch that design because the first thing you'll want to consider is the water, structures that will effectively gather water, and then the access. In some cases, this may involve changing the land to create canals or ditches to collect runoff.

Allowing nature to do what it does best

Have you ever watched someone doing a Sudoku and told them the answer to one of the squares? We think we are helping, but more likely, we have ruined their process. Permaculture is quite similar. We believe that our interference can help nature, and sometimes it does. But there are many occasions where permies have to take a step back and let nature do its amazing job. Chickens do a fantastic job of pest control while hunting and scratching for bugs. Nature has its own way of multi-functioning because chickens are, at the same time, preparing land for planting.

HOW PERMACULTURE GOES BEYOND ORGANIC FARMING AND GARDENING

To an extent, organic gardening and permaculture are very similar. Both are related to agriculture, not using chemicals, and making a positive impact on people's health and the environment. When I think about the two practices, I see organic gardening and farming as a business or hobby, whereas permaculture is a social movement. But the differences are a little more intricate than just my humble opinions. We will break down some of the core areas that separate one practice from the other.

Crops

When you go through your shopping list, there are certain foods, our staple foods, that are always on the list: bread, rice, pasta, and legumes. It's safe to say that these foods are in high demand and they aren't seasonal. They are, however, all annual crops, which we now know aren't the permie's best friend. Organic gardeners tend to focus on annuals, but because they require more maintenance, they may grow just a handful of different crops. Permaculture focuses on variety—you wouldn't want to eat just asparagus! Because perennials are low-maintenance compared with annuals, the workload is less for a permie, so they have more time to plant a greater variety with increased diversity. What's more, perennials can provide crops for years, if not decades.

Technique

Generally speaking, organic crops are planted in rows known as monocropping and may not use resources in the most efficient way. Monocropping goes against the ethics of permaculture, which we will discuss further on. On the contrary, permaculture ensures that every element benefits the next. Certain crops are planted together, each with its own function, whether that's to revitalize the soil or provide shade. Some plants can be used to attract a particular insect required for pest control. Rain can be collected for irrigating plants or to provide water for animals that also play their own crucial role in the ecosystem. The rule of thumb is that each element should serve at least three functions. An elderberry plant, for example, would provide shade in a garden, food for chickens, and a natural remedy to reduce fever in humans. Its hardwood can also be used as timber for heating and cooking. The elderberry plant complies with the multiple function principle.

Pest and weed control

It's not uncommon for organic farmers to use organic sprays to control pests and weeds. Again, if this is their source of income, it's understandable that they need to use some form of control to reap a bigger harvest. Pests and weeds have their rightful place in a permie's ecosystem. The use of any type of spray would create an imbalance, and it might kill the pests but it doesn't solve the problem as the pests will come back. If you have too many slugs in your garden, you can introduce a duck. Ducks love slugs, and their manure is excellent for growing plants and vegetables. Rabbits are happy munching on weeds, thus keeping weeds under control, and ladybugs eat the larvae of different insects that may otherwise cause havoc.

Soil

The soil we use is literally the food for our food, and we want it to be the best possible. Organic gardeners tend to till the soil once a year, and tilling disturbs the life within the soil. In permaculture, we tend to stick to either a dig-once method, a no-till method, or a chop and drop method. These methods build the soil by adding organic material to the top. The organic material is used as a mulch, feeding the soil and retaining moisture within it. These methods do not disrupt the intricate soil biology and instead benefit the soil while allowing the natural ecosystems within the soil to thrive.

THE SOCIAL ASPECT OF PERMACULTURE

Permaculture is a lifestyle that incorporates the home, garden, and community. It requires us to be mindful of the environment around us. I love the natural world because nature gets along with nature. Your mint might try and take over your chives, but the two aren't going to create a war that affects the rest of the garden. Humans don't respect relationships like nature does. Appreciating relationships in nature is an excellent way to encourage people to look at relationships differently. It begins by asking ourselves: What will happen to this existing relationship if I add another function or structure? What good will it do for my environment?

Fortunately, permaculture is beginning to positively impact society as a whole. When it comes to city planning, architecture, and even economics, we are now starting to consider what will happen to social interactions when certain physical structures are introduced. Thanks to permaculture, cities, towns, and neighborhoods are paying attention to the ethics and principles that can start healing the world rather than searching for the greatest profits.

Finally, society needs more diversity. Diversity leads to challenging outdated beliefs, new ideas and innovation, and strong cross-culture relationships. Permaculture shows us just how incredible diversity is, and it's helping the agricultural industry to appreciate different perspectives. Nobody has the right to judge a farmer for using chemicals, just as the farmer can't dismiss chemical-free farming. Working together enables the opposite opinions to see things from different points of view, leading to a shared understanding.

The principles of permaculture have their own impact on society but deserve a chapter of their own. The core philosophy and ethics of permaculture will also change how society interacts, grows, and transforms.

UNDERSTANDING THE PHILOSOPHY BEHIND PERMACULTURE

What nicely ties the social aspect of permaculture and ethics is the philosophy. The main philosophy of permaculture is working with nature rather than against it. It's to create human communities that design sustainable ecosystems by copying what works in natural ecosystems. It combines a closed system that supports the theory that no man is an island but instead a set of relationships and connections between components for the greater good of people, the environment, and the planet as a whole.

THE ETHICS OF PERMACULTURE

> *Science brings society to the next level; ethics keep us there.*"
>
> *Dr. Hal Simeroth*

Our ethics are codes that we live by that help us choose the right thing to do instead of the wrong. It's easy to put on a show and choose the right options when people are watching us, but it's our ethics that keep us on the right path when nobody is watching. Ethics can spring from our culture, background, the law, our society, etc. Permaculture wouldn't work if people didn't follow the primary ethics behind it.

Initially, there were three ethics behind permaculture, but we have included a fourth that is becoming more popular.

Care for the Earth

Earth creates and sustains all humans. Without Earth, there simply wouldn't be existence. It's in our own best interest to care for the Earth, not just for the air we breathe and the water we drink but for our children too. When we care for the Earth, we have to consider living and non-living things because there are relationships between all. The plants and animals we eat require healthy soil, the soil requires water that is not polluted, and so on.

Care for the people

Traditionally and still today, one of the harshest punishments to receive is exile, banishment from one's community, or solitary confinement. Humans are social beings, and although we don't always get it right, we need social interactions for improved mental health. It's important to remember that sustainability is not the same as self-sufficiency. Permaculture isn't about growing everything you need for yourself because we still need other things to survive, and we rely on our community for these things. What if your neighbor has eggs from chickens and you have potatoes? Wouldn't it be better for both of you if you traded products?

Fair share/reinvest surplus

I am reminded of the Buddhist proverb, "When is more enough." We have come to a point that even when we have food, shelter, and modern comforts, we still want more: a bigger car, a better home, a fridge full of food that ends up being wasted. When we only take from nature what is enough, nature can feed us for free. This ethic ties in nicely with care for the people. If you do grow more than you need and there isn't anything to trade, give to those who are less fortunate. This reminds me of a project by Hubbub in the UK. There are now 150 community fridges in the UK, a fridge where people can share their surplus food (Community Fridges, n.d.).

Animal care

Earth care encompasses wildlife, but if you choose to introduce animals to your ecosystem, it's only ethical that you care for them in the right way. We will take a look at the impact of different animals further on. However, to encourage the right balance and maintain an ethical permaculture garden space, your animals need access to food, water, shelter, and sufficient space.

So, permaculture in a nutshell? It's a social movement that is still relatively young but combines philosophy and ethics in order to create a sustainable ecosystem that mimics one that can be found in nature. It takes care of people, animals, and the planet without leading to greed. It's a hobby, a passion, and a lifestyle that is based on design and the use of natural resources. Then there are the twelve fundamental principles of permaculture that we will focus on in the next chapter.

Chapter 2

Principles of Permaculture

As we saw in Chapter 1, permaculture is more than simply naturally growing your food. It's a way of life that combines ethics and principles to develop an ecosystem that supports sustainability. Once we have discussed the twelve principles of permaculture, we will be able to appreciate the full benefits of this social movement.

The Twelve Permaculture Principles

1. Observe and interact

The very first principle isn't action but rather to observe and without preconceptions. Planning garden space shouldn't be about what plants look nice in a certain spot or planting vegetables closer to the kitchen. By meaningfully observing, we can see nature working at its best and how we can copy it so that each element goes where it will flourish. Observe the different interactions in your garden. Remember that you will have to observe the interactions throughout all four seasons and collect as much information as you can in order to maximize your design.

· · ·

2. Collect and store energy

If every home in California had solar panels, the state would generate approximately three-quarters of its energy from solar power (Köchel, 2019). But for those who don't live in such a sunny place in the world, there are other ways to use natural resources, for example, wind and hydropower. Water tanks collect and store water, and root cellars can be made to store food without using power. Linking to the ethic, Care for the Earth, the principle isn't limited to collecting and storing energy but also finding ways to reduce the energy we use. If you can grow ivy on your house, you can naturally keep your home cooler. Indeed, renewable resources will often require investment now, but bear in mind that this resource will soon benefit you and future generations.

3. Obtain a yield

Obtaining a yield in the eyes of permaculture is all about a delicate balance. We want future generations to enjoy a wide range of foods that we grow, but we also need to grow food that we can enjoy in the early days of our project. Furthermore, when we see results earlier on, the positive rewards encourage us to not only continue but expand our permaculture practices. There is also the balance of growing enough food for ourselves and hopefully others in the community without draining our ecosystem's natural resources. The great thing about yield is that, after careful observation, "yield is only limited by the creativity of the designer" (Mollison, 1988).

4. Apply self-regulation and learn from feedback

In permaculture, we have positive and negative feedback. When the conditions of your ecosystem are just right, plants grow, and you can enjoy the rewards: positive feedback. If a resource is consumed too quickly, nature will take over and adjust the system accordingly and may produce more or less: negative feedback. Our design should reduce the amount of negative feedback that may lead to inopportune growth. Positive feedback, however, speeds development. Ideally, the feedback from our garden will allow us to make small changes to work toward self-regulating systems, so there is little need for harsh corrective interventions from us.

5. Use renewable resources and services

Using non-renewable resources is like withdrawing money from your bank account without having a regular source of income. One day, there will be no money, and it will be too late to get that money back. The biggest problem with the world today is that if we can't physically see the bank account level, we just keep making withdrawals. Permies will always favor biological solutions, thinking about how a natural resource can do a task without consuming non-renewable resources. This principle shows us how to take passive benefits from our elements without destroying the entire host. An apple tree not only provides fruit but also shade. The wood of most fruit trees burns well, so you can use the wood for fires to cook with. Animals provide us with brilliant renewable services, especially for the preparation and treatment of soil.

6. Do not produce waste

There is no waste in nature. Waste generated from one element turns into something useful for another. The design of your permaculture space needs to minimize both pollution and waste. Through observation, find ways to use any abundance and waste that is produced. When grass grows too quickly, there is a risk of bushfires. Rather than using gasoline to power a lawnmower, a permaculturist would introduce grazing animals to overcome the abundance. Most of our garden waste like bark, leaves, branches, and twigs, can all be used to create compost and, later, humus. Outside of the garden, this principle also extends to other things we buy. We can use reusable bags for shopping and consider if the products we buy are ethical and, above all, really necessary.

7. Design from patterns to details

A common permie mistake is to begin with the details and over-look the bigger picture of the area they want to design. Every system has a set of patterns. By taking a step back and observing these patterns, you can see things from a holistic perspective— the tiny, intricate interconnects that make something whole. It can be confusing trying to work out the design details, so starting with the patterns helps to organize the details. It also prevents the planning of huge projects that look pretty but don't carry out the correct functions. Anyone can implement a complex design when they appreciate that it comprises various simple, functioning systems. To help with the design of space, we create zones and sectors. We plan our zones based on how often we use a particular element and how often it needs to be serviced. Sector planning is about the energies and elements of nature that pass through the area. Chapter 3 will go into more detail on zones and sectors.

8. Integrate rather than segregate — stack functions

It is just as important to look at the connections between the elements in a system as the elements themselves. There will be flowers and vegetables in your garden, and in the past, you might have been tempted to plant them separately to be more aesthetically pleasing. But, when you plant them together, you will find that pests are reduced. Try to have a minimum of three functions per element. You will find that the more closely your elements are connected, the the better the system becomes at self-regulating itself. To achieve this level of integration and self-regulation, all plants, animals, and infrastructure must be meaningfully placed. When you integrate wild herbs into grazing land for livestock, it increases biodiversity, mineral content of forage increases, and soil quality improves.

9. Use small, slow solutions

Have you ever been in a situation where someone has come in and made massive changes with the hope of making improvements but only succeeding in creating a disaster? The metaphorical bull in a china shop has no place in the permaculture garden. We do need to take on board the feedback from our systems, but we can't rush to make the changes. It is better to find small, slow solutions that are more efficient. Just because something grows fast, doesn't mean it's the best for your environment. It might make you feel like you are doing something right, but big and fast can often cause more damage than good.

10. Use and value diversity

When a child learns to speak more than one language, they aren't just able to communicate with more people. They get to absorb new cultures, learn more values, and understand different perspectives. They have the ability to develop more connections in the world. Our gardens should be the same. Monocultures are at a higher risk of damage from pests. Polycultures have a greater diversity of plants and crops, each bringing its own advantages and natural pest controls. With a more diverse range of plants, your household can enjoy more fruits, vegetables, and herbs. What's more, if you put all of your efforts into just one or two crops, if something goes wrong, you risk losing more. If you have twenty crops and something goes wrong with one or two, you still have eighteen or nineteen to help feed your family.

11. Use edge zones and value the marginal

There is a beautiful natural phenomenon that occurs at edge zones. An edge zone is where two different environments meet. For our example, we will use a forest and grazing land. The forest attracts a unique bird species that lives off elements in the forest; the same can be said with the grazing land. The edge zone combines both the birds from the forest and the grazing land and an additional type of bird that thrives in both zones. Land that is considered to have little to no agricultural or industrial value is called marginal land. If you look around your city, you can probably see land that can be put to better use. When people fear edge zones because they don't want to mix one element with another, they create more marginal land. Exciting things happen in edge zones and they should be part of our planning and design.

12. Creatively use and respond to change

Principle 9 looks at how we make small, slow changes to achieve more. The final principle shows us how it is better to use changes to our advantage. After all, nature will give us some changes that we weren't expecting, but we have to be creative enough to turn them into another advantage. We look at stone and water and assume the stone is stronger. But water has a way of making its way through the stone. We can fight change, often using wasted energy, or we can become flexible and adapt to the change. The systems we create have to be flexible to face change. A structure that can't change won't survive and, therefore, isn't permanent.

THE BENEFITS OF PERMACULTURE

It's crazy to think that there can be more benefits to permaculture even after what we have seen so far. We will look at some more benefits to this practice, not just for ourselves but also for our community, the Earth, and future generations.

- Permaculture isn't limited to traditional gardens. It can be practiced everywhere, from apartments and window boxes to community spaces and educational institutions.
- You can save massive amounts of money, not just on food bills but also electric and water bills and gardening resources.
- Permies have lower carbon footprints. They don't use chemicals, always choose renewable resources, reduce contamination and work toward a pollution-free environment in the air and the ground.
- Your green thumb becomes green living. Your food is chemical-free, you can learn how to use things that you grow as natural dyes, the way of living spreads into other areas of your life as you become more invested in taking your bike instead of the car. Your enthusiasm for green life is contagious and spreads throughout your neighborhood and community.
- You are providing animals with a safe haven. Aside

from deforestation, the world is seeing more and more wildfires that are wiping out entire ecosystems. These animals need homes or they can't survive.

- You are providing insects with a necessary home. Poor bees have a terrible reputation because of their survival instinct to sting, but we need them. Insects such as bees and butterflies pollinate around 35 percent of the world's food crops (Enjoli, 2020). If we are already struggling to feed the world's population, it's crucial we provide a home for those that are helping us.
- Permaculture improves air quality in urban areas. Humans need plants for oxygen, and the more plants we have, the better air quality we can appreciate.
- You and your family are healthier. The physical health improvements of eating homegrown, completely natural foods are well known. Growing plants also have a significant positive impact on your mental health. Green promotes a feeling of calm, lowers stress levels, and can motivate others. There is an immense sense of pride when a permie gets to watch their area flourish, which will also boost mental health.

INTERESTING FACTS THAT YOU MAY NOT KNOW ABOUT PERMACULTURE

In 2018, permaculture celebrated its 40th anniversary as a global practice. The early practitioners didn't need scientific evidence to convince themselves of the benefits. Today, we are somewhat skeptical and more so when the word "movement" is involved. Despite celebrating forty years of permaculture, researchers have been investigating the benefits of permaculture for approximately fifty years. These studies include climate research, the diversity of ecosystems, and the function of life from the soil, to name a few.

Perhaps when you first started reading this book, you thought permaculture was about a particular way of growing plants with a degree of planning. In reality, permaculture encompasses health; planning and decision-making; education; finance; engineering; and a humanistic, holistic mindset.

Detroit is a marvelous example of how permaculture can make life-changing differences. The town of Detroit was financially broke in 2013. There was no money, no opportunities, and people left in the thousands. Those who stayed adopted the permaculture principles and turned a nutrient desert into a new agricultural community producing thousands of tons of food annually.

The principles and ethics of permaculture don't encourage profit. That's not to say that permaculture can't be profitable. A traditional farm might use 12 feet for twelve rows of carrots. With permaculture, you can grow the same number of carrots and even some leeks and peas with less than 3 feet Take the profit out of the equation and you can still see how you can take advantage of small spaces to save money on food.

BUSTING THE COMMON PERMACULTURE MYTHS

Permaculture is not simply a gardening strategy that has been reinvented. We have seen that the practices have been around for generations and anything that lasts for thousands of years is something that we should listen to. What has happened is that scientific research has enabled permaculture to become a science backed layout of a system that supports a sustainable and regenerative way of living.

Organic gardening and permaculture are not the same, but we have already seen this. The myth behind permaculture is that it is organic growth in your garden. Permaculture makes the most use of everything that nature provides for us with the help of specific designs to maximize nature's gifts, such as food, energy, and shelter. At the same time, it follows the organic practice in that there is no use of chemicals.

PERMACULTURE AND THE FUTURE

Another great myth related to agriculture and humanity is that as soon as we developed agriculture rather than relying on hunting food, we became more civilized and less barbaric than our early ancestors. Permaculture professor Toby Hemenway pointed out that humans lived a sustainable life as hunter-gatherers for at least a million years. In 10,000 years, humans have made mistake after mistake in agriculture that has converted almost two-thirds of Earth's land into desert. One example is the Fertile Crescent of Mesopotamia. After 2,000 to 3,000 years of agriculture, the area still hasn't recovered, and that was 7,000 years ago.

Oil has helped us to rebuild some of the damaged soil with the use of machinery and technology. Modern agriculture still largely depends on oil. Logically, peak oil marks the beginning of peak agriculture, but what happens when all the oil runs out? It's up to permaculture to save humanity. I don't want to come across as dramatic or someone exaggerating the situation. You can't scare people into permaculture. Nevertheless, instead of farming, which continues to add to the problem, permaculture is sustainable and regenerative.

We need to turn our attention away from giant plantations that produce and then transport food worldwide. Instead, we need to start scaling down cities, making more green spaces for communities to turn into ecosystems that provide benefits for all. Your first goal should be to start in your garden, then encourage neighbors and others in the community to get involved.

Consider setting up a system where people can share or exchange abundance if this system hasn't already been implemented. . In just 3.7 acres of land, the Nordins in Malawi, produce more than 200 crops in a ten-month growing season (Aid & International Development Forum, 2018). Imagine a world where every community could create a garden that does the same!

I admit that, at first, I wanted to choose flowers, plants, fruits, and vegetables to grow, and things that I wanted to eat, see, and smell over something that would nurture a self-regulating environment. Between the ethics, principles, and myths, my biggest takeaway is that permaculture is about how you can help nature do its best in the area you have and not use nature solely for your advantage.

Now that you are fully aware of the principles and ethics of permaculture, it is that exciting moment when we can start looking at the actual design of your garden. The following chapter will look at design principles, our different zones and sectors, and ensure we incorporate the multiple functions rule of three for stacking elements.

CHAPTER 3

DESIGNING YOUR GARDEN—THE WHOLE SYSTEM APPROACH

YOU MIGHT BE FEELING that after the first two chapters, it takes more than just your average enthusiast to introduce permaculture to their garden. This might be true for farms, but you don't need to employ any specialists to get you going for home gardens.

Another thing to mention here is that if you already have an amazing garden, the last thing you want to do is flatten it all to start new. This chapter will look at designing your garden, whether it is a flat area or already well-established. Hopefully, after Chapter 1, you have begun the observation phase. Be patient! For those that have a blooming garden, observation is easy. For those who have just grass, you may not feel there is anything to observe, but trust me. There will be.

Back home on the farm, I made my first mistake. I observed the areas that already had crops and animals, and I neglected a space of land that we had always wanted to do something with but never had. On closer observation, I saw that at night, bats would flock to this area. I had never been a fan of bats until my professor explained that not only are they excellent at pest control, but they also disperse many seeds. Weeds were growing, so the soil conditions were right for growth, and after heavy rain, water collected at the edge of the land. Without my permaculture knowledge, I would have overlooked these tremendous potentials.

What is most frustrating is that you still can't get out there and plant. It's frustrating, but patience is a virtue and more so for the permaculturist. During this chapter, the good news is that you can get your pen and paper out, some colored pens, stickers, or even better, Post-it notes and Post-it stickers. I found that Post-its are ideal for permaculture planning because you can easily remove them and replace them if you come up with a better design idea.

While we look at what matters in our whole system approach, we will break down each element of the design. I know it's a little challenging, but once you get to the end of this chapter and you have your design, don't forget to go back and review the ethics and principles to ensure you have incorporated all of them. If you prefer, keep a brief list next to you as you are creating your design.

The Permaculture Zones in a Design

Some more complex permaculture designs can go up to nineteen zones, but generally speaking, designs go up to five zones. When you look at your zones, we plan from the center going outward from your home. Different elements go in each zone depending on the maintenance and how often they are used. If there is a side of your home that you can't access or you rarely go to, it wouldn't be included in the zones. The idea is that we optimize our energy use by keeping the things we frequently use close to the home. Now, let's take a look at how we could use each zone.

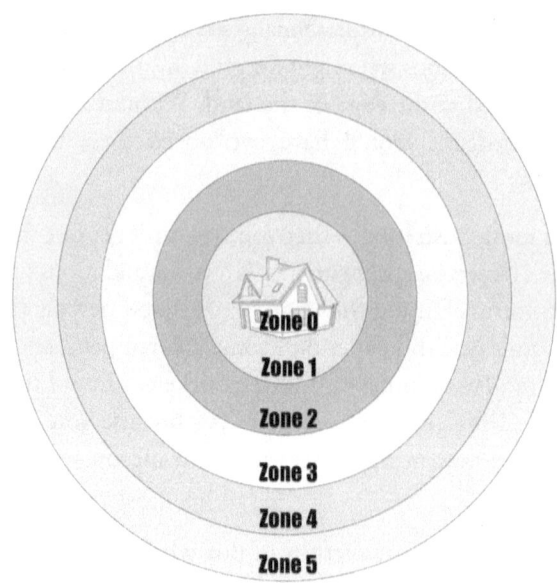

Image: Permaculture zone map

Zone 0: Your home or the center of human activity.

Zone 1: Plants, flowers, seedlings that require daily watering, log piles, compost, herbs for your kitchen, chickens for daily feeding and egg collecting.

Zone 2: Small fruit trees, berries, ponds, windbreakers, polytunnels, hardy perennials, and herbs. This area will still need watering but perhaps every other day. Also, consider herbs that reseed themselves and slow-growing vegetables. You can keep small livestock here too.

Zone 3: This zone is a good place for larger fruit trees, nut trees and is a good place for water storage. If you have a bigger space, it would be ideal for main crops and larger livestock like cows and sheep.

Zone 4: Hardy trees, unpruned trees, forestry, wild foods grown in nature, timber. It's a less-visited area that is only semi-managed.

Zone 5: Completely unmanaged, native plants and wildlife. There may be food that can be collected if there is plenty of it. The furthest zone from your home is for observation and learning.

There should be no strict borders between your zones because of principle 11 and the use of edges. Plants and animals at the edge of Zone 1 must be multifunctional so that the plants and animals at the edge of Zone 2 benefit.

STACKING FUNCTIONS

We have briefly touched on the rule of three and how each element should serve at least three functions. By ensuring elements have multiple functions, you are essentially outsourcing many tasks to nature. Two of the best examples are chickens and trees because they easily comply with the stacking functions rule of three.

Chickens can:

- Produce eggs
- Produce meat
- Produce feathers (compost or insulation)
- Control insects
- Control weeds
- Produce manure
- Ventilate and prepare the soil

Trees can:

- Provide shade
- Capture humidity
- Prevent soil erosion
- Produce food and oxygen
- Provide firewood and building materials
- Be used as windbreaks
- Help reduce the risks of frost
- Help control the difference between night and daytime temperatures

There is more to a permaculture garden than trees and chickens. Here are more ideas that will help you fulfill the rule of three in your design.

- Bees pollinate, provide us with honey, and promote biodiversity.
- Cows and sheep graze the land, fertilize the land, and provide meat and milk.
- Hedges can grow fruit, provide shelter and privacy, and become a habitat for insects and birds.
- Compost is a breeding ground for earthworms, a source of food for birds, and it's a soil amendment that is rich in nutrients.

Another good reason to use Post-its for planning is that you can include the different functions of each element on the Post-it. Then, if you do choose to move something around, you won't have to cross out or erase too much from your plan.

RELATIVE LOCATION IN A PERMACULTURE DESIGN

With multiple functions in mind, we can now look at the relative location of each element. Choosing the right location for certain aspects will also stop you from making numerous trips across the garden while taking advantage of the functions. Each element has inputs and outputs, and how we place them can strengthen their relationships by utilizing the inputs and outputs.

One great example is to build your chicken coup under messy trees such as peach and mulberry. The chickens do a great job of eating fallen fruit and fertilizing the trees while you don't have to worry about the mess on the ground. When planning relative location into your design, think about ways to reduce the repetitive tasks that you may not enjoy.

Plant tomatoes against your compost pile. The shade stops the compost from drying out, and the nutrients in the ground encourage delicious tomatoes. When emptying your compost bucket, you can pick tomatoes. Finally, when the tomato plant has finished, you can chop it up and throw it straight into the compost pile to feed the rest of the garden.

Deciduous trees can provide shade for a house in summer, and in winter, they allow sunlight through to warm the house. When you set up trellises, point them in a north-south orientation, this way, all of the plants can appreciate the sunlight. Any garden beds that don't smell particularly nice can have fragrant plants and flowers next to them, for example, lavender and Carol Mackie.

If you have a slope in your garden, design the area so that the water storage is up high on the slope. Gravity will take care of supplying the home or irrigation systems that are lower down. If you are going to introduce a pond, consider creating it to cross two zones, taking advantage of the diversity that can grow and be attracted to the different zones. A birdbath invites wildlife into your garden, many of which will help with insect control. Don't forget to put it in a line of sight so that you can see them from your window. After all, your garden still has to be a pleasure to look at.

PERMACULTURE TURNS PROBLEMS INTO SOLUTIONS

Sometimes, nature, our gardens, or a combination of both will present us with what appears to be a problem. Part of becoming an experienced permaculturist is to understand that not everything is positive or negative. When problems arise, and they will, we need to treat them as solutions to either solve another problem or make improvements in other areas.

Bill Mollison couldn't have put it better with one of his more famous quotes,

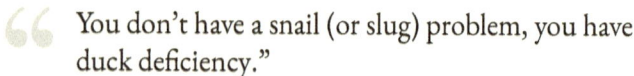 You don't have a snail (or slug) problem, you have a duck deficiency."

He found a solution for every surplus and deficiency, therefore removing problems from the permaculture garden.

If there is a spot in your garden that is always boggy, the grass won't grow because it is too wet. It's the perfect location for a pond but if you aren't ready for that step, allow for a bog garden. You can plant water chestnuts and lilies, and have an area for your ducks.

PERMACULTURE DESIGNING FOR DIVERSITY

Principle 9 introduced us to diversity and the benefits of polyculture. Greater diversity leads to more stability and resilience. As soon as you start mixing a greater variety of plants in one area, insects become confused and steer clear. Relationships between the different plants develop over time, continuously evolving.

You probably have an idea of the plants you want in your garden. Go back to your Post-it notes and think about what these plants require. There will be different nutritional needs, heights, light, and space requirements. Before finalizing what plants, flowers, herbs, and vegetables to mix, consider these points:

- Having too many of the same plants together makes it easier for pests to find them. It also makes it easier for diseases to spread. Caterpillars will struggle to find your lovely leafy greens if you have surrounded them with marigolds and chives.
- Companion planting is the concept of plant combos that do well when grown together. For example, flowering herbs pair well with squash, tomatoes, basil, lettuce, and garlic.
- Did you know that there are more than 10,000 tomato varieties? Some are better for salads, others for sauces. Planting different types of tomatoes increase diversity and can extend your tomato season.
- Combine deep-rooted vegetables (carrots) with shallow-rooted vegetables (radishes). Both require different nutrients from different parts of the soil.
- Be diverse with your natural pest controls. As we have seen that chickens clearly have multiple functions, so do birds and lizards. Lizards are very sensitive to pollutants and do a great job in helping manage insect populations. If your garden attracts lizards, you know there are low levels of pesticides (remember, even if you don't use them, it can take years for pesticides to disappear from your systems).

DESIGN SCALE—USE OF SPACE

After talking about ponds, rock gardens, and cows, you might be looking at your space and wondering how you will fit everything. Even if you don't have the biggest of gardens, there are still ways you can maximize the use of space. Some of these methods we have seen before, like planting radishes and carrots together; the same applies to carrots and onions. They don't need separate spaces because they pair well together. Not only do their roots grow differently, but their leaves too. Onion leaves shoot straight up, allowing light to reach the carrots.

In one small area, you can have herbs, creepers, and grasses as a bottom layer. Shrubs and small trees make up a middle layer, and climbers and tall trees make up the top layer. These layers are known as stacking. Each element has access to the necessary inputs, and the outputs benefit each layer.

Guild planting is a way permaculturists stack different plant species to get the most out of vertical spaces, light, and nutrients in the soil. It encompasses companion plants, so it helps to know which plants work well together and which don't. Before finalizing your design, check if your plants are going to get along with each other.

CHAPTER 4

Your Comprehensive Guide to Permaculture in 9 Steps

You might be surprised to know that there is still a considerable amount of flexibility in permaculture. There seem to be a lot of rules, principles, and things to remember, but these are more like a set of guidelines to make sure you incorporate the ethics and get the most out of your space and efforts.

While I am sure you are just itching to get out there and plant that first seed or introduce your first livestock, this chapter will tie together the final concepts to include in your design and some preparation before the all exciting implementation. Call it the cherry on the cake, the last bit of studying before you earn your permie wings!

This next chapter will cover the 4 Ps of permaculture, an understanding of regenerative agriculture, and finally, an overview of the nine steps that we will be examining in detail throughout the subsequent chapters.

SUMMARIZING PERMACULTURE WITH THE 4 PS

The 4 Ps stand for place, patterns, process, and principles. We have covered most, but it's nice to have a short recap to refresh and clarify.

1. Place

The place would seem obvious; it will be your garden, balcony, window box, etc. But the first place you should look to is you. What do you need and want in your place? Next, take a look at your surroundings and how elements can be added. Finally, take a look at your place from the street. Also, look around the community. Once you have looked at the bigger picture and what can be done for the community, bring it back to your place at home.

2. Patterns

Wherever you look in nature, you will find patterns. It could be in the flow of water, leaves, and flowers, even our DNA. We need to integrate our needs and wants into the patterns of nature instead of trying to work against them. These patterns that we can see in nature will work as a big advantage for us and our space.

3. Process

The process begins with our goals and observations. It is how we choose to design our permaculture area based on what we hope to achieve and what natural resources are at our disposal. The notes and planning we made in Chapter 3 are all part of the process.

4. Principles

So we aren't quite as overwhelmed by everything we need to incorporate principles. Heather Jo Flores uses a simplified version of the principles (Flores, 2019.). This version includes:

- Diversity/goals
- Patterns/observation
- Specificity/boundaries
- Recycling/resources
- Creativity/analysis
- Placement/design
- Inclusion/implementation
- Autonomy/maintenance
- Feedback/evaluation

Again, it might feel like there are a lot of rules dictating what can and can't be done in a permaculture ecosystem. But as long as you have ticked off all 4 Ps and you know in your heart that the ethics are being followed, the rest is down to your priorities. Not everyone is going to be able to grow carrots for the whole community, that's OK. Perhaps your priority is to have a beautiful, diverse range of flowers; that's good too. Maybe you will just start with some tomato seeds, make your own compost, and collect rainwater for the plants. These are all great starts.

Five years ago, I met Ben. He lived in an apartment with a small balcony, and this is exactly how he started, with one packet of tomato seeds and a vision. He gave some of his neighbors some tomatoes to try, and a few of them said he could grow other plants in parts of their gardens. He started to share these vegetables. When word got around about Ben's gardens, someone offered him a small plot of land that allowed him to take his permaculture to a larger scale. There are always possibilities that arise when you are using natural resources to help others and the planet!

AN ESSENCE OF REGENERATIVE AGRICULTURE

Regenerative agriculture is an amazing concept that joins the core agriculture practices we need: permaculture and farming. Regenerative agriculture puts soil at the heart of our systems to regenerate it to increase organic content and fertility. It is a global approach that enables soil, animals, and plants to create food chains between the three ecosystems. It is a holistic approach, and with time, commitment, and education, farmers will reap the same benefits of healthy soil as permaculturists do.

How does regenerative agriculture work?

The food industry is responsible for approximately 26 percent of greenhouse gas emissions (Poore & Nemecek, 2018). If we continue doing what we are doing at the moment, within fifty years, we won't have enough arable topsoil to feed the world's population. Regenerative agriculture not only does no harm to land but also improves soil quality. When we revitalize the soil, we can grow highly nutritious food, and on a global scale, we can even reverse climate change. The process involves rebuilding soil and organic matter and reviving degraded soil biodiversity. The benefits also include water retention and carbon draw-down, a process of capturing CO_2 from the atmosphere and storing it in long-term solutions such as plants, soils, and oceans.

What are the key techniques?

Less to no-tillage: Tillage is the mechanical agitation of soil to prepare land for crops. It releases large amounts of carbon dioxide because of the machinery used. Furthermore, it creates many disturbances within the soil, which is not good for soil microbes. When farmers reduce plowing and tillage, they can increase the soil quality and help reverse global warming through carbon drawdown. The top layer of soil is rich in carbon, disturbing it exposes it to oxygen and creates carbon dioxide. Without tilling, carbon remains in the soil.

Greater diversity: Richer soil makes it easier for farmers to plant a wider range of crops. Plants release different types of sugars into the soil via their roots. These sugars provide nutrition for microbes in the soil, and the microbes return the favor by giving plants a greater variety of nutrients. The result is a more diverse range of produce and a better yield.

Better use of fertilizers: Regenerative agriculturists are more cautious about chemical fertilizers because of the damage to the soil's health. Like permaculture, there is more emphasis on the natural relationship between plants and microorganisms.

Crop rotation: When farmers stick to planting the same crops in one area, the soil ends up with an abundance of certain nutrients and a shortage of others. Rotating crops ensures that the soil is exposed to more nutrients and prevents diseases and problems with pests.

Cover crops: When land is left with no crops, the elements will wash away nutrients or the land will dry up. A good rule to remember is that bare soil is bad soil. Cover crops are like living mulch. They slow down soil erosion and improve the availability of water. They can also help with disease and pest control.

What is the connection between regenerative agriculture and the climate?

All across the world, we are noticing more and more extreme weather conditions. California has suffered some of the most devastating wildfires in the last few years—more than ever before. Australia and Greece have had similar destruction and devastation. The other extreme is the floods in areas where significant floods were previously unheard of. European countries have not only lost lives but billions of dollars in damage to property. In London, 3 inches of rain fell in ninety minutes; the River Thames rose by 1.5 feet in less than five minutes. In Belgium, 10,000 residents had to move from their homes, and in Germany, 184 people lost their lives. These extreme weather phenomena caused by climate change are negatively impacting farming and the food industry.

Wildfires release carbon dioxide because of the carbon stored in trees and vegetation, and these fires also release other harmful gases such as methane and nitrous oxide. The more carbon released, the more the global temperatures increase and the more wildfires we have. The climate feedback loop goes on.

The water cycle is also negatively impacted by climate change. The warmer the plant gets, the more evaporation and, therefore, the more rain. But this rain isn't evenly distributed across the planet. Rain is the perfect example of natural irrigation, but when there is an extreme downpour, the rainwater runs off because the ground can't absorb that much water at once. Nutrients wash away, and plants drown, or there are severe droughts.

We can significantly impact the global climate crisis by strategically shaping land to prevent runoff, covering bare soil with plants and mulch to prevent evaporation, and building healthy soils that promote microbial life, water retention, and carbon sequestration. Controlling carbon emissions and restoring the health of topsoil is a global responsibility, and we can all do our part whether we have a small garden or a large farm.

As more and more farmers see the effects of climate change on their lands and livelihoods, more are making the positive change toward regenerative agriculture instead of waiting for science and technology to make radical changes. It makes sense for farmers to have more plants across their land. During photosynthesis, plants absorb carbon dioxide and release oxygen. But they also convert carbon dioxide into carbon. Some of the carbon is used for growth, and the rest is transported to the soil. Once in the soil, it becomes soil organic carbon and provides food for microbes and nutrients for the plants. And so, another loop is completed.

General Mills provides crops for more than one hundred brands in one hundred countries (General Mills, 2021). The company is one of the leaders in setting the example for regenerative agriculture. By 2030, they plan to implement regenerative agriculture on one million acres of farmland. What's more, the company is dedicated to helping other farmers switch to the same techniques.

As the soil is the heart of our ecosystems, there will be more information on regenerative agriculture. In Chapter 8 we will focus on soil health and fertility, and Chapters 8 and 9 will discuss a garden with no tilling. By the time we reach the end of Chapter 11, you will have a complete understanding of plant diversity, crop rotation, and cover crops.

The 9 Steps to Starting Your Permaculture Practice

What can happen at this stage is a psychological term known as analysis paralysis, and we have to be careful to avoid this. As the name suggests, we analyze a situation so much that we are paralyzed, and no action occurs. Between the ethics, principles, and 4 Ps, there are so many decisions that we can't finalize our design, or we start to think that maybe it would be better to start at a later date. This section will take us from start to finish with the nine steps to building a permaculture garden.

Without further ado, here are the nine steps to prevent analysis paralysis and turn your permaculture dreams into a reality.

1. Observe and record. Create a site map based on your observations of the patterns of your space, contours, water sources, and natural features you have to work around or structures you may have to remove. Take note of the sun's path and angle in winter and summer, wildlife, and other natural influences on your multi-layered site map.

2. Based on your observations, patterns, and the information in Chapter 3, you can design what elements will work best in each zone. Think about the bigger elements and then plan the details.

3. It's time to start sourcing your plants and materials based on the inputs and outputs of each. You may want to use pots to plant seeds so that you have some seedlings for when your soil is ready. Other seed types can be planted straight into the ground.

4. The water system is crucial for your garden. Rainwater harvesting might be a barrel under the gutter but it can also be far more sophisticated. Your water system needs to prevent runoff and a wasted resource and instead, spread the water around the garden.

5. Soil preparation will involve testing for quality and making sure there is the right balance. It's important to know how to naturally adjust mulch and compost to create healthy soil. For example, eggshells add calcium, fruits and vegetables provide vitamins, and limestone reduces acidity.

6. The first thing to plant will be your anchor plants within the guild, or pattern. These will be the primary plants of the companions you have chosen. We plant them first so they have time to anchor their roots well into the soil. It will also give them extra time to gain some height before their faster-growing companions are planted.

7. Next, it will be time to add your companion guilds. These could be your pollinators or herbs for pest control. Before planting the companions, double-check the companion guide and the inputs and outputs of each.
8. Even though the design has been created, you will still need to maintain your space. You can build your compost pile, start to harvest herbs and spices from the garden as teas, and of course, continue observing your system for imbalances. It will be helpful to create a calendar and to-do lists to keep track of tasks.
9. Remember, you will see positive feedback and negative feedback. The negative feedback allows you to work with nature and make minor adjustments to find a better solution.

A permie can have the strongest wheelbarrow, the sharpest hoe, or the finest quality seeds. But really, a permie's greatest tool is their patience. You have done very well to look outside and just imagine what could be. Your patience will be rewarded as you reap what you sow! And the first reward is to start the next chapter and get familiar with your soil and surroundings.

CHAPTER 5

BIOMIMICRY—OBSERVE AND RECORD THE LANDSCAPE

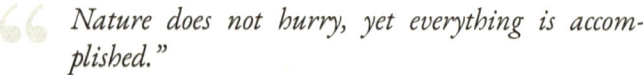 *Nature does not hurry, yet everything is accomplished."*

Lao Tzu

I HAVE LEARNED SO MUCH over the years as a permie but one of the greatest lessons has been patience. I love this Lao Tzu quote. I have been where you are, and I know what it's like to go out and buy all the right tools and start planting something, anything, just to feel like I am making progress. Put the trowel down! We must start with observation. Look outside and remember that nature, unlike humans, is in no hurry. It listens, learns, and adapts. If you are struggling with patience, ask yourself if you want one pea-sized tomato for your salad, or do you want so many tomatoes that you can make tomato sauce for the entire neighborhood?

Think back to some of the greatest inventors throughout history. They didn't get their inspiration from a laptop. They watched how nature made the best of every detail and designed mind-blowing inventions for their time. These are the fundamentals of biomimicry.

WHAT IS BIOMIMICRY?

Biomimicry takes patterns and designs we see in nature and uses them to produce materials, structures, and systems. The most obvious example of biomimicry is Leonardo da Vinci's flying machine that would allow him to "soar like a bird." Engineers use the streamlined shape of the bird to design planes today, and they also consider how birds behave to fly better. Birds use their beaks to groom their feathers, ensuring their body is smooth, and planes copy the same smooth design. If that's not enough, here are some other amazing examples of biomimicry:

- **Spider web glass**: Car windshields use a special design taken from a spider's web. This way, the glass cracks but doesn't break.
- **Wind turbines**: A humpback whale has ridges along its pectoral fin that create an aerodynamic flow in water. This is how wind turbine blades got their design.
- **Traffic**: By studying the way ants move in specific rows, we can better understand how to manage traffic on highways.

My personal favorite is VELCRO®. I don't know how many times it has come in handy. It wasn't until my dogs and I were covered in burrs that I discovered that George de Mestral had noticed the same thing. The tiny hooks of the burrs were the inspiration for the product. You may not notice a natural pattern that will lead to a worldwide invention, but you will be able to appreciate what your space has to offer.

WHAT CAN BIOMIMICRY TEACH US ABOUT OUR SPACE?

Biomimicry in permaculture looks at nine concepts that we need to bear in mind when observing. From nature, we can learn:

- **Everything needs sunlight:** There are only a handful of known species on the planet that don't need sunlight. With stacking, we can take advantage of space but ensure all our elements have adequate light.
- **Nature only uses the energy it needs**: Even when there is an abundance of something, nothing gets wasted. Squirrels can bury their nuts and perhaps forget where they buried them, and even foxes bury what they kill to dig up later. If the food isn't eaten, it either sprouts a new plant or decays and becomes part of the soil.
- **Functional design comes from nature**: Sharks can't chew, so they have expanding jaws and incredibly sharp teeth to shred or eat prey whole. Sea lions have flippers for swimming but claws to help them move on land. In both cases, the form of the animal fits the function.
- **Everything in nature is recycled**: From energy to matter, everything that exists in nature gets reused. There is no waste.
- **Nature supports cooperation**: No man is an island in nature. Plants and animals rely on other plants and animals to create productive working relationships.
- **Diversity is vital for nature**: In each ecosystem, you can find a diverse range of creatures with different needs, yet they successfully cohabitate.
- **Nature requires local expertise**: You will often find that, in an ecosystem, there are species that are considered to be the local experts.

- **Nature can control excess**: If there aren't the right amount of resources for a species, nature can slow down the growth of the population. Without the right nutrients, some species are less fertile, reducing a drain on the already limited resources.
- **Nature understands limits**: Something, unfortunately, humans don't. We don't respect the planet's limits, and we are always trying to take more than we give.

It's clear to see the link between the concepts of nature and biomimicry within the principles of permaculture. It's amazing to see how energy is used in nature the same way we try to use energy in permaculture. Diversity is key and beneficial relationships are required. Based on biomimicry, there are certain patterns that you need to start looking out for.

THE PATTERNS OF LIFE

If you don't know where the problem is, you can't find a solution. We have already mentioned that there aren't problems in permaculture, only positive and negative feedback. Once we can put our green thumb on this feedback, we can find the solutions for our space.

Soil quality is essential. You will want to look at the type of soil you have as well as the drainage available. PH levels are going to have a huge impact on growth. A pH reading of 1 to 5 is going to be highly acidic to very acidic. From 8.5 to 14, the soil will be very alkaline to highly alkaline. The ideal range is 6.2 to 7, but your climate and plants will impact this.

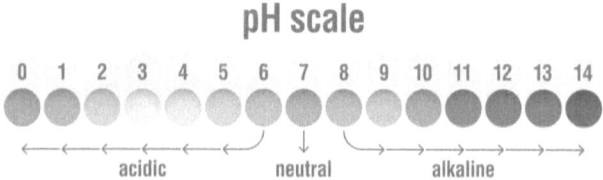

Image: pH Scale

On that note, what type of climate do you have, and what are the weather patterns like? Think back over the last few years and see if you have noticed any changes to weather patterns, such as longer, drier summers or more snow than usual. Look at different areas of your space and decide how much sun, shade, and wind each section gets. Remember to consider these elements for each season.

Take time to observe what is thriving in your garden. Even if it just seems like weeds, we can find a way to use this to our advantage by learning what type of weeds they are. Do you have any wildlife like particular insects, birds, frogs, or small mammals?

For successful observation, you need both a routine and a particular spot to sit and observe. Choose a spot and try to get there by making the least amount of noise and disruption. It's worth trying to spend at least fifteen minutes, even half an hour, sitting very still in your spot. It can often take that long for any wildlife to come back after your initial appearance. Be sure to return to your spot at different times of the day. Sitting in the sunshine sounds blissfully relaxing, but you will have to observe in the rain too. The same goes for frosty mornings and windy days.

Never assume that something isn't relevant. As all things in your area have a relationship with other things, every observation should be written down. In the long run, you will save time and money by focusing on the details at this stage.

How to Survey a Site

After the necessary time observing, you should end up with a decent pile of notes, perhaps even a full diary or notebook. Don't rely on your memory. It would be easy to confuse insect activity at sunset in autumn with bird activity at sunrise in early winter if it has been months since your observation. If you rely on your memory, you risk making a mistake, which means the time you have spent observing could be wasted.

Line graphs are an excellent way to plot things like rainfall over time. A map is an excellent resource because it makes it easier to visualize your space. That being said, you probably aren't going to find space for all your observations on one map. Create a base map with the permanent features, buildings, areas that are too difficult or expensive to change, and things that are already working well. Then, use tracing paper as an overlay to record other information. If you are technically inclined, multi-layered digital maps are an excellent choice and can be made on platforms such as SketchUp Pro, CAD, or Google Maps. Let's look at some methods to organize our observations.

PASTE

The acronym PASTE helps us to identify the main elements of our space. These include:

- **P**lants—trees, herbs, weeds
- **A**nimals—wildlife, birds, insects
- **S**tructures—permanent features that aren't going to be moved
- **T**ools—objects that have a job to do like solar panels

- **E**vents— human activity, natural activity

There isn't a way to add an F and still make an easy acronym to remember, but we can't forget fungi. You may have to be creative with your words like it takes a Fun-guy to paste a plan or something equally cheesy.

DAFOR

DAFOR helps us to keep track of plant life over the seasons. You will already have a list of plants, but you should add more information like **D**ominant, **A**bundant, **F**requent, **O**ccasional, or **R**are. You can use DAFOR to monitor animal activity too, just remember you are better off looking for evidence of animals rather than the animals themselves. You have more chances of seeing a fox's footprints than the fox.

Desire lines

These are the paths or routes people and animals frequently take or make. We may have a path from the front door to the street or the garden to the tool shed. Needless to say, animals won't always choose the same path as humans. Look for exits through hedges or under fences.

Zones and sectors

We have already looked at zones; however, they aren't always going to fit into nice little rings around your home. For example, Zone 1 covers elements you care for or frequently use, including your desire lines. Your sectors will be determined by the wild energies such as the sun and wind. Both of these energies are directional, so be sure to mark on your overlay map the directions that they come from.

Topographical sectors

Unlike directional sectors, topographical sectors are made by the landscape. To understand topographical sectors, we must get out there in extreme weather conditions. When it's raining heavily, can you see any rainfall down slopes or areas of erosion caused by the rain? Where does the rain from the roof drain, and are the gutters all intact? Check to see if there is any runoff caused by the rain. How frost settles can tell you about trapped cold air and where your frost-free areas are. Take advantage of the snow to discover your desire lines.

Rocks

Interestingly, rocks can give you more information about the landscape than you may think, and even rocks have their own cycle of creation, erosion, and recycling back into the system. The types of rocks you have will impact the soil and, therefore, your plants. Some rocks can boost soil conditions, and others can make them worse. Rocks can be porous and hold a lot of water or be impermeable.

Leaks

Any form of leak can be seen as a wasted resource. Look for heat escaping from your home or other buildings. Leaks in pipes and guttering need to be patched, so you don't waste any water before it can be harvested and taken advantage of. Large grass areas can be seen as a wasted resource because they are high maintenance with little to no output. As you are observing, make a note of any resource that can be put to better use.

Utilities

It is worth having a separate overlay to mark your utilities. Use different colors to mark phone lines, gas, electricity, water, and sewage. If you are going to do major landscaping, you need to make sure you don't damage any of these structures. Contact your local utility companies to help determine the correct locations of the different utilities.

Getting into the observation habit now will help you keep up with the practice as your permaculture space develops. For every small change you make, you need to observe the changes and record the results. Before and after note-taking will be like your personal handbook specific to your space. A gift to pass on to the kids so that they can continue the sustainable lifestyle.

With your observations, maps, and overlays, it's time to start deciding exactly where you want your elements to go.

Chapter 6

From Patterns to Details—
Designing

About three years ago, I was called to help a guy with his 2-acre garden. He had implemented a few elements but quickly realized that he had skipped the design phase. For example, his beautiful raised beds only had about a 12-inch gap in between. This was great because he had paid attention to the principles and wanted to maximize land use. But he couldn't get a wheelbarrow down the middle, which made an awful lot of extra work. Design in permaculture matters!

It's understandable why we have such a sense of urgency when considering how much the planet needs our help. But to rush any stage leads to flaws in the design and more work in the long run. Imagine installing a water storage system that fills up too quickly, only to have excess rainwater draining away—or removing the poison ivy but forgetting that the roots can still hurt. So, it's still not quite time to break in the new trowel, but each bit of patience gets us a step closer.

A good permaculture design provides a framework, knowledge, and tools to start the process with fewer regrets and mistakes. A design helps us to interact with nature, even when it throws us something unexpected. It allows us to choose the right things for the best output while considering the important connections between the elements.

How to Put Your Observations into Design

There are a few more acronyms that help you take everything you have learned so far and create your design. While I can give you all the advice in the world, I can't outline the design because, as you can imagine, there is an infinite number of designs depending on where you are in the world and what you want to achieve. Even if you don't know what you want to plant yet, you will find more inspiration on this later on. For now, let's look at how to design your space based on your observations.

A framework that comes from landscape architecture is SADIMET, and each letter is a different stage of the process.

- **S**urvey
- **A**nalysis
- **D**ecisions
- **I**mplementation
- **M**aintenance
- **E**valuation
- **T**weaking

A similar framework comes from industrial engineering. OBREDIMET has slightly more detail in the observation stages:

- **O**bservation
- **B**oundaries
- **R**esources
- **E**valuation
- **D**esign/Decision

- **I**mplementation
- **M**aintenance
- **E**valuation
- **T**weaking

Finally, a more simplified framework might be a good reminder for those with a little more experience who can remember the details required by each of the CEAP steps.

- **C**ollect site information
- **E**valuate your information
- **A**pply the permaculture principles
- **P**lan a schedule for implementation, maintenance, and feedback

My personal favorite is GOBRADIME because it's easier to remember and encompasses the right amount of detail for the design.

- **G**oals: Your goals must be SMARTER (specific, measurable, achievable, relevant, timebound, ecological, and rewarding).
- **O**bservation: All that we learned in Chapter 5.
- **B**oundaries: Visible and invisible, find or create boundaries, change them, or establish them.
- **R**esources: What resources does your space have and what does it need? Look at the abundance of resources your space has.
- **A**nalysis: Your goals combined with everything you have got up until now can be analyzed for the best outcomes. Be careful not to overwhelm yourself and suffer from analysis paralysis.
- **D**esign: Maps, overlays, looking at the patterns of nature to design your elements. The design must be

logical and, of course, realistic.

- **I**mplementation: Careful planning of design elements into to-do lists and onto calendars.
- **M**anagement: The 4 Ms of management are monitoring, maintenance, messes, and mistakes. It's the learning stage.
- **E**valuate: Go back to your observation spot and evaluate your progress and any changes you want to make. Take time to enjoy your achievements.

Whether you are tweaking or evaluating, the fun doesn't end here. Each of these frameworks will become a continuous cycle. Every time we evaluate the feedback, it's time to go back to the pen and paper, decide on our new goals in line with the principles of permaculture, and design the next elements we want to introduce.

OPTIMIZING YOUR PERMACULTURE DESIGN

There are many aspects of the design, such as our zones, sectors, desire lines, and plant companions, and we have to consider the location of each element. We have probably all made the mistake of impulse buying a plant or flower because it looked nice but got it home and realized that you don't have the optimal location for it. If we look first at the location, we can make the right choices when selecting and purchasing plants. Other things that you need to work into your design are elevation planning, soil types, integration, and of course, you have to visualize how your space will grow.

Elevation planning

Even gravity can be seen as a resource. You can use things like a hydraulic ram to pump water uphill, but doesn't it make more sense to elevate the water tank and let gravity do the hard work?

Another example would be to put a solar hot water panel under the hot water storage tank because you can benefit from the thermosyphon effect instead of a pump.

Soil type

You may have different soil types in your space, which you will have discovered during the observation. From your zones and sectors, you might have planned for certain plants in certain zones but now realize that the soil type in that zone isn't suitable for the plants you wanted. It's worth taking time to modify soil conditions so that they are optimal for the plants you would like in each zone. This is especially important in Zone 1.

Visualize the future

The tree you plant now won't provide much shade, but what happens in five or ten years? The area around your tree will have more shade in the future than it does now. The design we are making now is a snapshot of the present, not the future. Be sure to think about any changing conditions that may impact your garden in the future.

Integration

We briefly looked at the importance of integration rather than segregation. Neat little borders are cute but they don't mimic nature. To integrate, you can look at the energy cycles available and link as many of your elements into the energy cycles as you can, such as solar power to provide electricity for your home and heat to a greenhouse if necessary. Remember your edges, as these will be where the most interactions take place. Nature rarely creates straight edges, and you will find that borders and edges in nature are curved to take advantage of more edges, greater diversity, and integration.

What are process flows?

In an ideal world, we will create all that we need from our own space and the natural resources available. Still, this won't always be the case. If your soil isn't up to planting standards, you may consider bringing in manure to prepare the soil. In return, you are able to grow more fruits and vegetables to give away or sell if this is going to be your income.

This is known as process flows, creating trades with the outside world to bring in additional resources you need to give back to the community. When there is any type of exchange with others, you need to incorporate this into your design. Any resource that comes in or goes out should have easy access to avoid wasting your energy that could be better spent elsewhere.

6 ESSENTIAL MAPS FOR YOUR DESIGN

Most people are happy enough to draw their maps, but you can use a computer if you feel a bit tech-savvy. You might have to collect certain maps from local authorities, such as your utilities' layout. For elevation planning, you can probably find a contour map of your area online or again from local authorities. Next, it's time to work on your maps. The base map will be on paper or cardstock, and the other maps can be on tracing paper.

- **The base map**: The outline of your site, the permanent features. Be sure to mark north and your size scale.
- **The sun map**: Highlight the areas of your garden that get full sun, part sun, and shade. You might want to do a sun map for summer and winter.
- **The sector map**: The things you don't have much control over but can use to your advantage. Some examples include contour, wind, wildlife, snow/fire, views, etc.

- **The zone map**: Organizing the relative location of elements in the design based on the level of use and maintenance needed. Previously discussed in Chapter 3.
- **The water map**: Where the rain falls and is collected and what needs irrigation. Are you going to have a pond or swales?
- **The master map**: From using the different overlays on your base plan, you can create your final master plan. It's a big decision, so you may want to make a master plan A and B to narrow down the best ideas and then combine the two to get the final master plan.

Know Your Inputs and Outputs

This is where things get fun! Write a list of every element you want in your garden or space. You might want to create a chart to keep things organized. For each element, you need to come up with all of the inputs and outputs. Think of this as NFY: needs, functions, and yields. The needs or inputs are all the requirements of each element to grow and thrive in the environment. The functions will be the outputs or the results of our inputs, including the materials it produces. With our needs and functions, we should obtain both a primary yield (what we had planned for) and a secondary yield (something not expected), but we can tweak our plan to use it as an advantage. See the elements inputs and outputs chart as an example.

Next, you need to find a way for all of your elements to match inputs and outputs, as we have seen before with the trees and chickens. It's unlikely that what you want in your ecosystem will connect every input to an output. When elements don't connect, consider adding extra elements to create the necessary balance and fill any gaps.

Element	Input	Output
Potatoes	Full sun, slightly acidic soil, loose soil	Potatoes, leaves that humans can't eat, food for another animal such as rabbits or compost
Ducks	Grain, insects, water	Eggs, insect control, meat, feathers
Lemon Tree	Well-drained, sandy soil, pH between 6 and 7.5, shaping once or twice a year.	Fruit rich in vitamin C for food and drinks, shade, leaves for salads, and soups.
Bees	Water, flowers, shelter	Pollination, help plants grow, honey and wax
Mint	Moist soil, sun, a little mulch	Herb for cooking, sauces, tea, pest deterrent, leaves that can improve skin tone

Image: Example of elements inputs and outputs list

"Sowing" It All Together

At this point, you are going to have a solid idea of what is going where. It's very close to the moment when we start to implement the design. There are just a few more things to incorporate, not necessarily relating to the design but bringing it all together without feeling like your design is getting too much for you to handle. There is an order of establishment that will help you appreciate the bigger picture of your design in smaller steps.

1. **Water**: All of your water elements need to go first, including a watershed, swales, water collection, ponds, etc.
2. **Access**: Paths and driveways to your house and the different zones, your desire lines, and fences.
3. **Structures**: Now that the correct access has been created, you can build any structures you want to

include. This could be anything from chicken coops to sheds for storage.

4. **Soil**: Preparing the soil, neutralizing pH, mulching, and compost systems.
5. **Planting**: With healthy, prepared soil, it's time to get those first plants in.
6. **Animals**: If you have the financial resources, you can add plants and animals simultaneously, but plants should be the priority.

These steps will help you to create a plan of action for your design, but there are five more tips that will help you implement your permaculture plan like a pro:

- Be big and bold with your design but don't feel like you have to implement everything immediately. You can start small and add elements of your design in stages.
- Remember that if you go big straight away, you will need the time to maintain...big! Think about the scale of your project, and be sure to implement only what you have time to take care of.
- Think of your critical elements first. Water and soil are the best starting points but decide what is essential to your design and prioritize those elements.
- Consider modular designs for large areas. Modular designs enable you to repeat units or guilds across the bigger space.
- Plant or build the bigger elements first, then add the details around them. This is essential for stacked elements.

After all this, you might be questioning your design capabilities. But you don't have to be a professional designer in any way to design your permaculture garden. You have created so many things already without overthinking them. Every time you plate food for dinner, you work out what looks right and what food should not touch each other. In your home, you design the best layout for your furniture and make changes as you see better uses of space. The same will happen in your garden. This initial stage may be daunting, but you will start to find confidence. Practice over and over again with your paper and overlays. Make mistakes here before even stepping foot in the garden center. In fact, don't be tempted by the garden center until you have finished Chapter 11, our plant guide.

CHAPTER SEVEN (PART 1): YOUR ECOSYSTEM'S LIFELINE—WATER

I HAVE this very fond memory of my cousins rushing around the house and garden watering plants before my grandparents arrived. It wasn't an inspection, more like an unconscious look at my aunt's plants with a few "gentle" tips to keep them alive. She took the hint when they only gave her succulents rather than risking anything more high maintenance. The fact is, fruits, vegetables, and flowering plants won't last more than around four to seven days without water.

It hurts to see rainwater washing away down drains only to have people then turn their garden hoses on. In this chapter, we will learn all about capturing and channeling water to make the most of this natural resource.

HARVESTING RAINWATER

Rainwater harvesting (RWH) is collecting rainwater to reuse as drinking water, water for livestock, or irrigating crops. The uses for rainwater are quite extensive, and you can use it to fill up swimming pools, fountains, and ponds, and it's perfect for washing your pets and cars too. Rather than letting rainwater wash off and go to waste, we can store and recycle it, and at the same time, reduce one of our utility bills.

Needless to say, it's not just about saving money. RWH can help prevent soil erosion while keeping your garden irrigated with water you can collect from your roof. Each time someone starts to collect rainwater, we reduce the strain on the world's drinking water system. Cleaning and purifying water for nearly eight billion people requires massive amounts of energy and generates huge amounts of pollution.

After reading the previous chapters, you will now appreciate that harvesting your rainwater is both ethical and complies with the principles of permaculture. Before we look at the techniques of RWH, let's look at some rules and guidelines to help master the fundamentals.

Observe thoroughly

You guessed it! Permies can't get enough observation. When observing water patterns, look at where your water flows and the different paths it takes. Look at places where it collects and where it is running away.

Start at the top and work your way down

If we want to take advantage of gravity, we need to start at the top. Find the highest point in your garden or space. If you have a contour map, you can easily spot it. If not, you may have to explore a little. Follow the flow of water up to the highest point.

Start small and simple

Think about infiltrating water into the soil. Is one giant change going to be more effective than multiple little ones? Making one big change can cause more problems to the ecosystem than helping, so it is always better to start small and monitor feedback.

Spread and infiltrate

Water will always take the path of least resistance. After a lot of rain, you will probably observe little streams of water making their way across the land, sadly eroding soil as it moves. The problem is that there is more rain than can naturally infiltrate the soil. RWH allows for more infiltration spots so water doesn't run off but can be usefully spread across the site.

Always plan for overflow

It's a bit challenging because you can't predict exactly how much rainwater storage you need. During observation, you might have a rough idea of monthly rainfall throughout the year, but one heavy storm can cause rainwater to destroy your beautiful beds and erode the soil.

Maximize living, organic groundcover

Groundcover plants, for example, clovers, are shallow-rooted plants. These plants act as a living sponge because they absorb large amounts of water and help protect your beds from runoff. Another function of groundcover plants is that they are high in nitrogen and boost soil quality.

Stack functions to maximize relationships and efficiency

We have seen how we can stack plants so that each layer has access to sunlight. We can do the same to take advantage of water and relationships. Groundcover plants are sponges, rich in nitrogen, and are edible. Ponds create an ecosystem of their own, as well as bio-swimming pools.

Continually reassess the feedback loops

Every RWH technique you introduce into your garden needs to be monitored to understand the impact on the ecosystem. There will be positive and negative feedback, both of which you can learn from and make small changes to see further improvements.

HOW MUCH RAINWATER CAN YOU HARVEST?

There are going to be variables that influence how much rainwater you can harvest. Someone living in Hawaii is going to have a lot more rain than someone in Arizona or Texas. Remember to consider the land and surfaces you have. Here are four questions to ask yourself to determine your RWH potential.

How many watersheds do you have?

The watershed is all of the lands that water and other dissolved materials flow through by gravity. To understand your watershed, start from the highest point in your land and watch where it flows and where it leaves. You may have more than one watershed. One may start at the top of a slope, and others may begin on your rooftop and down through the gutters. Knowing where your watersheds are and how much water is flowing is essential, but you also need to think about the surface they run on and the speed.

Have you mapped the flow?

You can use your contour overlay map to mark the flow of your watersheds. Arrows can indicate the direction, but you can also use bigger arrows for greater flow and even longer arrows for faster flow. Remember that when you place the overlay on the base map, the flows should correspond with the placing of your buildings.

Have you calculated the total rainfall?

Total rainfall is an average figure, and you should always plan for overflow just in case. There are various ways to do this. You can calculate per month, per season, or per year. Work out the square footage of your area, say a roof that is 10 feet by 12 feet. The total area would be 120 square feet. Multiply the total square footage by the rainfall in feet. 120 square feet x 2 feet annual rainfall = 240 square feet annual rainfall. Finally, multiply the total square feet of annual rainfall by 7.48 (gallons per square foot), and you have the total gallons of rainfall for that area. 240 square feet annual rainfall x 7.48 = 1,795.2 gallons rainfall per year.

Have you calculated runoff volumes?

The runoff from a metal roof is going to be higher than the runoff from grass. Here are some figures to help you understand how much runoff you can expect from different surfaces:

- **Bare earth**: between 35–55 percent
- **Grass**: between 10–25 percent
- **Sand**: 30–50 percent
- **Asphalt and concrete**: 80–90 percent
- **Metal**: 95 percent

After calculating your total rainfall, multiply it by the surface type to better analyze your rainwater harvest potential. Using the figures from the example above, let's factor and calculate runoff:

1,795.2 gallons rainfall x 0.95 (metal) = 1,705.44 gallons rainfall per year from a metal roof

Rainwater Harvesting Techniques

There are some rainwater harvesting techniques that we are going to include in later chapters. These techniques will include choosing sensible plants to do the work for us (in Chapter 10) and using proper mulching techniques (like sheet mulching in Chapter 8). For now, we will start this section by looking at one of the most popular rainwater harvesting techniques.

Swales

A swale consists of a trench or basin and a berm. Swales help to slow, store, spread and sink the water. It's like a trench that can capture water and prevent runoff. Based on your observations, you will already have a good idea of swale placement, but there are a few other rules to bear in mind. Keep your swale 10 feet away from buildings and drain away from the building. Swales should also be 18 feet from the edge of a steep slope or a septic drain field. Plan your swale to be uphill from the garden. You should also perform an infiltration test, so you know the soil will absorb a minimum of 1 inch per hour.

Image: Vegetative swale cross-section

For larger swales, you will need to use contour levels to mark your contour lines, along with flags every 6 feet or so. Trenches are normally between 6 and 30 inches deep and 1 foot to 4 feet wide. As you dig the trench, be sure to create a mound (or berm) on the downhill side. The angle of the slope also needs to be measured. If it is too steep, there could be a chance of blowout from the build-up of water, overflow, or the soil could be damaged. Your slope ratio should be no more than 1:3 rise/run or 33% percent slope. If you are going to create steeper slopes, you should think about adding terraces to your design. Finally, use a level to make sure the trench bottom is even.

To test your swale, watch what happens the next time you get some heavy rain. If it overflows, you know that you will have to dig deeper or wider to accommodate more rainfall. Once the size is right, you can plant things like perennials, hedgerows, shrubs with berries, or fruit tree guilds.

Check-log Terraces

Check-log terraces are short walls on hillsides that slow down water runoff, and they are a great way to create a hillside garden with loose, aerated soil. These systems become their own little ecosystem with air and water tunnels made by earthworms and beneficial fungi around the eroding logs.

Image: Check-log terrace

With your contour map close by, drive stakes along an elevation contour line. Remember, you want to make sure that your contour line will capture the rainwater on the flatbed. Your stakes should be 2-6 feet apart. If your slope is quite steep, keep the stakes closer together. Starting from the bottom of the slope dig out a terrace so that the area between each terrace is level. Set aside any topsoil that was removed and save it for the top level of the terrace.

Next, you need to lay logs or limbs on the uphill side of the stakes. Again, the size of your check-log terrace will impact the choice of logs or limbs. 1 to 4 inches in diameter is the ideal size. If you choose anything over 6 inches, you might have to dig a trench (then lay cardboard) so that the log sits in the trench and does not push too much weight against the stakes. Your logs and limbs want to be piled slightly higher than level as, over time, they will settle.

Use twigs and leaves to line the uphill side of your logs. You want to use these leaves and twigs to plug any holes. These will also stop the soil from filtering through your logs. Finally, add your soil to the terrace garden space and expect to add more later as the soil settles. When you build the next level, add the topsoil that is removed to the previous layer. Before planting anything, wait for heavy rainfall to check for any areas that might need more logs, leaves/twig plugs, or soil. Once the top level of the terrace is complete, bring the topsoil that was removed from the first level and add it to the highest level of the terrace.

French drains

A French drain lets you take problematic buildups of water and redirect the water to areas that need it. If you imagine your water flowing down a slope, you might build a trench across to catch the runoff and allow it to slowly infiltrate the soil. Or, you could install a French drain to take the water to your flower beds or vegetable patch.

To incorporate French drains, find the collection point where rainfall tends to sit and mark a path to the area where you want the water to end up. Be sure to check your plans to ensure you will not dig next to and damage existing pipes or utility lines. The trench you dig will be approximately 6 inches wide and between 18 and 24 inches deep. Also, remember that gravity has to work its magic, so your trenches will have to get gradually deeper. The general rule for this is at least a 2 percent drop or one-quarter inch drop per foot.

Line the trench with landscape fabric. The fabric will stop small objects from getting into the pipe but will allow water to pass. Adding landscape fabric to this project is optional and will depend on your specific situation. For example, if you plan to put dirt or plants on top of the drain, it's probably wise to use the fabric. If the drain is filled with only rocks, the fabric might not be necessary. If using fabric, leave at least 10 inches of fabric on each side of the trench because when you add the rocks or gravel, it will be pulled down. On top of the fabric, place about 3 inches of gravel. Next, place a perforated pipe onto the rocks with the holes facing **down**. You might assume that more water would be absorbed with the holes facing up but remember that water will always take the shortest path to the lowest level, so you will capture more water with the holes facing down.

Fill the trench with gravel or rocks, and then wrap the remaining fabric around the gravel. What you cover the pipe with is up to you. Some people will cover it with soil and groundcover, and others will add more gravel. Just remember, if there is ever a problem, the more you put on top of a french drain, the more you will have to remove.

There are a few other rainwater harvesting ideas that you may want to consider depending on the size and the design of your garden. I have briefly included some other ideas and how they work.

Infiltration basin

An infiltration basin is a sunken area that collects stormwater. The water is stored in the depression and slowly filters into the soil. Infiltration basins are good because they are cheap to make and can be used in small and large areas. As the water filters through the soil, pollutants are removed. However, they aren't the best for areas of high water pollution. You can combine infiltration basins with trenches.

Sunken beds

As you might imagine, sunken beds are the opposite of raised beds but with more advantages. They can offer protection from the wind, mulch stays nicely in place, and the four berms allow for water capture. My advice would be to warn any visitors what you are planning with your sunken beds. In the preparation phase, they can look like empty shallow graves!

Waffle gardens

Waffle gardens are sunken beds grouped in the shape of a waffle. They have all the benefits of sunken beds, but you do have to consider your design more. If you aren't going to leave a small path to reach the middle of the "waffle," you wouldn't want to plant anything that requires regular visiting.

Making the Most of Recycled Plastic for Drum Rainwater Tanks

For a beginner permaculturist or someone who doesn't have that much space, you might find that the above options are too much. Others like to have a simple solution in Zone 0 and Zone 1 to collect water for hand-watering things closer to home. For this, we will use (mainly) everyday objects and recycled plastic drums to harvest rainwater.

You will need:

- The appropriate number of drums for your calculated runoff (remember to plan for excess and overflow)
- Concrete blocks
- Flexible downspout connector
- 4-inch atrium grate or 4-inch net cup
- Fine mesh fabric, old nylon tights work well
- 0.75-inch rain barrel spigot kit
- Corrugated black plastic pipe (1–1.5 inch diameter)

Steps to create your drum rainwater tanks:

- Place the concrete blocks under the gutter downspout or drain where you collect water from the roof of your building.
- You'll want to set up the drums on concrete blocks because raising the drums will improve gravity and increase water flow.
- Cut a hole in the lid of the drum, wide enough to fit your atrium grate. Attach the fine mesh to the atrium grate to make a fine mesh debris filter that sits on top of the hole. The filter will require regular checking for debris build-up and cleaning if needed.

- Attach your gutter downspout to a flexible downspout connector to direct rainwater into the hole with the filter.
- Drill a hole at the bottom of your drum and attach the spigot kit. Attach your garden hose or fill up your watering can from the spigot you created at the bottom of your drum.

In most houses, one drum will not be enough to collect all the water runoff. To account for overflow, it would make sense to attach a second or even third drum, depending on the size of your drums and calculated runoff.

To attach a second drum to collect overflow, it's quite simple. Cut a hole in the main drum and the same size hole in a second drum at the top of the drums. Connect the two holes with a corrugated black plastic pipe. When the primary drum is full, the overflow will pass into the second drum.

ARE YOU READY TO BUILD A RAIN GARDEN?

Rain gardens are another lovely solution for different sized areas, and they don't take much effort. A rain garden is a shallow depression with a berm on the downhill side. Rainwater that falls on impermeable surfaces can be directed to the rain garden. To make the most of your drum's collection system, you could direct your overflow to the rain garden.

Image: Rain Garden

Once you have discovered the ideal location for your rain garden, calculate how big it needs to be. Measure the runoff area that water is being directed from in square feet and multiply this number by 0.08 to get the volume in cubic feet. For example: 800 square feet multiplied by 0.08 equals 64 cubic feet. Next, divide this by 1.1 to get the surface area in square feet. So, 64 cubic feet divided 1.1 equals 58.18 square feet surface area for the rain garden.

Now you know how many square feet to mark as your area to dig. Dig out 12 inches of soil. Set aside the topsoil for when it's time to refill the rain garden and use the subsoil to create the berm on the downhill side. It's important to level and then aerate the bottom of the bowl you have dug out. The subsoil needs to be broken up and have some holes dug into it for proper filtration. As always, you must check that the rainwater collected will effectively infiltrate the subsoil. Fill the bowl and wait for twenty-four hours. The bowl should empty within this time. If it doesn't, you can aerate the subsoil more or lower the height of the berm.

The rich topsoil that you dug up from the bowl can now be added to the bowl. You need to cover the base with about 6 to 9 inches. If you don't have enough, top it off with compost. Finally, you can decorate your rain garden with native perennials and deep-rooted prairie plants. Deep-rooted plants are essential because they soak up excess water in the wet season, and in the dry season, their long roots have access to water deeper down in the soil. Add plenty of mulch to your new plants to keep the surface moist when it's not raining.

How Can You Tell a Rain Garden from a Pond?

The main difference is that the amount of water in a rain garden will fluctuate. There will always be water in a pond, which creates a different ecosystem that is often rich in wildlife, such as frogs and dragonflies. Ponds will have different depths with shallow areas providing foraging and warmer water. To create a pond in your garden, you can use the following tips:

- Choose a location that receives a good amount of sunlight but with some shade areas during the hotter months.
- Avoid planning for a pond under trees. You will have more work come autumn when leaves fall.
- Choose an area in your garden that is already waterlogged or a place where there is a natural dip. Preferably you'll want to choose the lowest point of your space.
- Your pond should be deeper in the middle with gently

sloping sides. Avoid straight lines. Curves increase the edges you have in your pond.

- Plastic membranes are popular for ponds, but natural resources such as clay and manure are preferred in permaculture.
- When choosing plants, be sure not to introduce anything that will take over the local environment.
- Include plants that provide a good balance, such as rooted and floating plants, marginal plants, oxygenating and submerged plants.
- Always check local guidelines and with local authorities before installing a pond because permitting is often required.

Harvesting rainwater should be seen as our obligation to help the planet. It may not be until you experience a burst or frozen pipe that you realize how much we take our faucets for granted. There are plenty of ideas to harvest rainwater depending on the size and design of your garden. The two key takeaways are to make sure you calculate your runoff and be sure to have an overflow system.

Chapter Seven (Part 2): Channeling Life through Your Garden

To continue with the importance of water and making the most of this natural resource, we will take a quick look at irrigation systems for your permaculture garden. While Part 1 looked at harvesting and storing water for your garden, Part 2 of this chapter will look at irrigation systems that help us passively direct water across the garden so that we can water every plant in the garden easily.

At this point, you might not have your beds set up, but that's not a problem. It's sensible to plan for irrigation while you are designing the rainwater harvesting systems. That way, you know that implementation, when the beds are ready, will be flawless. It also helps to get all of your water system materials at the same time.

Clay Pot Irrigation

Clay pots have been used for thousands of years. They originated in Northern Africa but made it to the States with the Spanish, hence the name Olla, meaning pot in Spanish. These clay pots are unglazed, so they are porous. There are plenty of shapes and sizes, all of which look elegant and natural. The traditional design has a large rounded body with a small neck.

Clay pots are buried in the ground, and only the neck or the opening is visible. Fill the clay pots with harvested rainwater. The water then seeps into the ground and reaches the roots of your plants. Clay pots conserve water because the soil and roots only absorb the moisture they need. It is estimated that the clay pot system can use up to 90 percent less water than other irrigation systems. Clay pot irrigation is an excellent solution for dry climates.

Another great advantage is that because you are only watering the areas that you want, you can reduce weed growth. Other irrigation methods include irrigating the entire area, but clay pots have more control over the soil you wish to irrigate. You could include clay pots near trees or even inside pots. To make your garden more efficient, you can set up self-filling systems with float valves, so you don't need to hand-fill the pots.

When deciding on the location of your clay pots, stick to the rule of thumb that one and a half gallons will provide water seepage for an 18-inch radius. With this in mind, you can plant around clay pots in circular forms 18 inches from the neck. Tall-growing plants don't need the clay pot so close, and can be 3 to 5 feet from the plant. If you have vines, you can place your clay pot approximately 9 feet away .

DRIP IRRIGATION SYSTEM

Drip irrigation is a super simple system that can go underground or sit on top of your beds. The central part of the drip irrigation system is a long pipe with holes in it. Water leaves the holes, and as it touches the soil, the droplets spread in a cone shape. There are different variations to buy, or you can purchase poly pipes and make your own holes that align with your design.

There are two main types of drip irrigation systems. You will need to choose from pressure compensating and non-pressure compensating drippers. Your choice will depend on the inclination of your garden. Non-pressure compensating drippers are more suitable for flat areas and pressure compensating drippers are best used on slopes, so you don't end up with more water at the bottom than at the top.

With your base map and an overlay, draw the lines where you want your drip irrigation system to go. In most cases, they are long stripes, but you can use spiral designs too. Plan how many feet of pipe you need (add 20 percent, so you have wiggle room if your plan changes). You will also need T connectors, elbows, and end caps.

Attach all of your parts and lay them on the ground where you want them to go. Flush the pipes out with water to ensure there isn't any soil or dirt potentially blocking the drips. Finally, add the end caps so there is no wasted water.

A slightly different design is to have one main larger pipe running through your garden and thinner pipes coming off. The thinner pipes can lead straight to the plants that you want to receive water. You can add different-sized dripper heads depending on how much water you want to be released on the end of the pipe. The typical dripper heads come in 0.5 gallon, 1 gallon, or 2 gallons per hour. It's best to use the same size dripper head for all your garden. If you have plants that need more water, you can add another pipe.

You might choose to bury or anchor your drip irrigation system. If you do, be sure about the placement of the pipes and holes. If you leave the pipes above ground, it will be easier to make adjustments if necessary.

Many people will attach their drip irrigation system to a garden tap. Yes, it's easier, but it's not the permie way! If you use drums to harvest rainwater, you can attach the drums with valves to your drip irrigation system.

GRAYWATER MANAGEMENT SYSTEM

Graywater management systems can range from relatively simple to somewhat more complex, not necessarily in how we create them but in knowing how to set them up and use the graywater correctly. Before we get into the details of using graywater, let's look at what graywater is.

Graywater is appropriately named after its color. Graywater is water that has already been used in the home and can be repurposed for irrigation. Graywater excludes toilet water but does include water used for laundry, hand washing, showering, and bathing. Just because the color isn't the most appealing, it doesn't mean our garden won't reap surprising benefits from graywater.

So, how can this "dirty water" be used to water our gardens? Water that has been used in the home contains small traces of dirt, food, grease, and hair. What we consider pollutants to water become food for plants providing nutrients and fertilizers, and it's actually better for them than clean water.

There are certain things that you can't use with your graywater system. These include dishwashing detergents and certain soaps. Your cleaning products need to be free from salt (sodium), boron (borax), and chlorine bleach. You should also try to make sure you are using pH-neutral products to not upset the balance of the soil. For example, liquid body gels will have little impact on the soil, but some bar soaps can lower the acidity of the water. Graywater systems aren't complicated, but you do need to pay more attention to the cleaning products you use.

Whole house graywater system

A whole house graywater system takes all of the water outlets from your home and connects them to one pipe with a 3-way valve. This valve allows you to direct the graywater to a surge tank or the sewer. In the surge tank, there is a pump that pushes the water out of a separate pipe to irrigate the garden. You will need a pump if your garden is uphill from your graywater outlet.

Laundry to landscape

The laundry to landscape is a much simpler system that you can set up yourself without making any adjustments to your plumbing. It does require your garden to be lower than your washing machine so that gravity does the work unless you are going to install a pump. All you need to do is attach your washing machine outlet to your irrigation system. You can also choose to only take the graywater from your showers if your house's foundation allows access to the drain pipes.

How to use graywater effectively

The ideal setup is to use graywater in conjunction with rainwater harvesting. This way, you can flush your garden with rainwater and give the plants and soil a break from graywater. The flush will lower the risk of potential toxic accumulations in the soil. It's important that none of your graywater runs off onto neighboring properties. Lastly, not all plants like graywater, and you shouldn't let any graywater come into contact with the edible parts of a plant.

Plants that grow well with graywater and those that don't

While you are planning which plants will go where (again, more in Chapter 11), you might want to consider having certain zones that will predominantly use graywater. Planning graywater zones prevents plants that are not suitable for graywater from coming into contact with it.

The following lists will give you an idea of which plants to irrigate with graywater and those you can't. Things you can irrigate with graywater include:

- Most fruit trees
- Raspberries
- Blackberries
- Gooseberries
- Rhubarb
- Passion fruit
- Grapes

- Oaks
- Honeysuckle
- Roses
- Juniper
- Bearded Iris
- Butterfly Iris
- Spotted gum

- Swamp Oak
- Weeping Bottlebrush
- Bird of Paradise

An idea of some plants that don't take well to graywater include:

- Ferns
- Rainforest plants
- Root vegetables
- Herbs

- Camellias
- Rhododendrons
- Agaves
- Aloes
- Sedums

- Foxgloves
- Primrose
- Deodar Cedar
- Bleeding Hearts

There are just a few other considerations that are important when using graywater in your garden:

- Always check your state laws before implementing a system, especially when changing your home plumbing system. You don't usually need a construction permit for a laundry to landscape system, but it's still worth asking your local authorities. If you are altering your plumbing system, you may benefit from using a licensed plumber in case rebates are available or permitting is required. My advice is to consult someone with graywater experience if you are at all unsure.

- Never store graywater for more than twenty-four hours. Nutrients start to break down and there might be a nasty smell.
- Don't let the graywater pool. Bacteria can build up, and this may be dangerous. It can also encourage mosquitos. Only irrigate as much as necessary.
- If your graywater isn't going underground, be sure that it goes under mulch.
- Keep pets and children away from graywater and wash your hands after coming in contact with it.
- Keep a close eye on soil pH levels. Listen to the feedback from your garden to make any necessary changes.

Despite the extra care, graywater can save you a lot of money on your water bill and help the environment. Not only are we not wasting water, but we are also reusing water that would otherwise end up in the sewer or septic tank.

Chapter 7's Parts 1 and 2 together provide many ways to harvest rainwater and use it to irrigate your garden without costing a fortune. While these solutions tend to focus on garden spaces, don't forget that you can still use small drip designs for your balcony window boxes. For example, you can poke small holes in a plastic bottle and plant the bottle in your window boxes. There are also clay spikes that are small enough for window boxes and they look nice too.

CHAPTER 8

A Love Affair with Soil

HEALTHY SOIL DOES a lot more than just provide our plants with all of the necessary nutrients to grow. Soil provides people with essential nutrients for survival through food, and it absorbs carbon, which helps regulate the Earth's climate. Taking care of our soil makes it more resilient to drought, floods, and fire—extremes we are witnessing more frequently. In this chapter, we will learn how to take care of the soil so that soil can return the favor.

What Can Nature Teach Us About Soil

Take a look at an undisturbed area of nature to see the soil at its richest. Look at how forests have a wealth of growth at all levels, from grasses to vines meandering around tall trees. Fruit falls from trees, animals eat the fruit; in return, their manure fertilizes the soil. Leaves and twigs fall, break down, and return to the soil. Never does a person get involved and till the land. Nature doesn't get a meterstick out to measure the distance for seeds.

What happens above the ground helps to preserve the topsoil. By preserving the topsoil, life below the surface has a chance to thrive. Microorganisms and insects are allowed to do their job, providing nutrients for the world above the topsoil, without disruptions. New beds are continuously being created on top of the old, with no digging! Digging is OK if we need to shape the land, but it isn't a long-term solution because the soil doesn't get the chance to replenish itself. We want to mimic nature as closely as possible to encourage the growth of the richest soils.

ALL YOU NEED TO KNOW ABOUT COMPOST

To create the fertile soil that our permaculture garden needs, we have various solutions. One of the simplest, which also helps reduce our waste, is a compost pile. A compost pile is a way to stack all of your organic waste so that they decompose. You can add all types of garden waste, from grass cuttings and leaves to kitchen scraps. The more variety you have in your compost pile, the more the microscopic organisms will eat away and produce more nutrients.

For the best compost, try to keep a ratio of 80 percent carbon and 20 percent nitrogen materials. Your carbon materials are brown and dead, for example, dried leaves and grass you cut some time ago. Nitrogen materials are green cuttings, manure, and kitchen scraps. Also, be sure to keep it aerated, meaning air gets into the pile. If you notice that your compost pile smells bad, it may need more carbon and needs to be turned to aerate. The more often you rotate your compost, the faster it will be ready to use on your beds. Compost needs water so that it is slightly moist. Avoid excessive amounts of water. A good rule is if you squeeze a handful of the compost, a couple of drops should drip out. If a stream of water runs out, then your compost is too wet. When the compost is a dark, earthy color with a rich smell of the forest, you know it's ready. We will look at compost piles more in improving the quality of the soil.

If you aren't keen on using a plastic bin, there are other ways to create a compost pile. Here are four other ideas:

1. A compost heap. Just an area in the corner of the garden where all organic matter gets tipped. It's not always ideal because you can't easily access the bottom, nutrient-rich compost without disturbing the top layer of matter.

2. A round wire bin. Sixteen-gauge wire fencing works well, and keeping it between 48 and 60 inches will keep it stable. Once the material has turned into compost, you can tip it over and stand it up again.

3. A wood/pallet three-bin system with three sides wood (or three sides pallet) and a wire door to hold the compost in each bin. The three-bin system makes it easy to turn and keep the compost aerated. Three bins are ideal because you will always have compost available. Begin by adding food scraps to the first bin on the left, every time you add food scraps, add some leaves too. When it is full, you can move the contents into the second bin and start filling up the first bin again. When the first bin is full the second time, move the contents of the second bin into the third and the first into the second. Now, new matter can go into the first bin, the middle bin is resting, and your third bin has finished composting.

4. A compost heap in a day. Find an area where you want to use compost. Add a layer of sticks and twigs. Add a second layer of carbon materials: dried leaves, grass, or straw. The third layer should be nitrogen materials: green grass cuttings, food scraps, and coffee grounds.

The last layer is a layer of soil. Place all of the layers on top of each other on the same day. You won't need to turn it and in two to three months the heap will be ready for planting.

Taking Advantage of Soil's Best Diggers: Worms

The official name for worm composting is vermicomposting. Worms are a great way to accelerate the process of turning organic material into an organic fertilizer known as worm castings. Worm castings are a super food for your plants and soil because they contain all the essential nutrients for plants, they increase the soil's water retention, and they add an unfathomable amount of beneficial microbes to the soil. In addition, worm castings contain humic acid which helps increase nutrient absorption in plants.

If you want to have a continuous supply of fresh worm casting all you have to do is get an old plastic bin, add worms with organic material (they love kitchen scraps, shredded paper, cardboard, leaves) and let them do their job. Their tunneling helps to create natural airflow through the material so it won't need turning. Just make sure your bin has holes in the bottom for the proper drainage. You don't need any particular type of worm, but California Red Worms aren't particularly long, and they breed very quickly. When it comes to removing your new vermicompost, feed the worms on one side of the bin. After a few days, they will have migrated towards new food, and you can harvest your worm casting without harming them.

Alternatively, you can use two stacking bins. In the base of the top bin, drill around fifty holes with a 1/4-inch drill bit. Around the top of the bins, use a 1/8-inch drill bit for ventilation holes all the way around. Add shredded paper, your organic matter, and worms to one bin and top with more shredded paper, spray to moisten, and place a lid on the top bin. To add more scraps, lift off the shredded paper, add scraps, and replace paper. When the vermicompost is ready, add the second box on top of the compost. Again, add shredded paper and scraps. Place the lid on top. What will happen is that the worms will migrate through the holes into the top bin, and you are left with worm-free compost in the original bin.

If you have an old bathtub, you can turn it into a neat worm farm. Place the bathtub on some concrete blocks to use the drain as an outlet for excess liquid. Layer the bottom of the tub with coarse gravel, then cover the gravel and sides of the tub with a cloth liner. The next layer is the worm bedding. Again, shredded newspaper works well. The food scraps will be added on top of the worm bedding. Like the bins, add shredded paper as a blanket to keep the environment moist and dark. You will need to cover the entire tub with a sheet of timber for protection from critters. As with the first bin, feed the worms at one end to harvest the finished compost.

You can feed your worms fruit and vegetable scraps, bread, cooked veggies, grains, coffee grounds, tea bags, and eggshells. Only add small amounts of citrus and onions, and never add meat, fish, or dairy. If you are adding shredded paper, make sure it isn't glossy or bleached.

How to Create a No-Dig Garden

No-dig gardening or lasagna gardening is a way to reduce work and let nature do its job. In many ways, no-dig and no-till are the same. Rather than churning soil to mix added nutrients, materials are added on top of the soil, and nutrients seep into the ground without disturbing life in the soil.

No-tilling is a favorite in permaculture. The Fukuoka no-tilling method was created by a Japanese farmer called Masanobu Fukuoka. His system involved planting rice. When the rice was close to harvest, he planted barley seeds in the rice field. Once he had harvested the rice, he laid the rice straw back on top of the field which provided a thick layer of mulch for the barley to grow in. You might not be growing crops to the same extent as Fukuoka, but no-tilling and adding straw is an excellent way to create soil.

No-tilling is also the basis of Elaine Ingham's healthy, alive soil. The biologist focused on bringing the soil back to its natural state so that it can produce its own fertility. Her approach incorporated an entire soil food web, where beneficial bacteria, fungi, protozoa, arthropods, and nematodes from aerated soil and the correct compost techniques led to a healthy soil ecosystem.

Making a no-till/no-dig garden is very simple. You can choose to make raised beds, a section of your garden, or the entire space, and you only need a handful of materials. You can mark out the area with chalk or wooden pegs or add your raised bed edging. There will be more on raised beds in the next chapter.

1. The ground needs to be prepared depending on the type of surface:

- Soil: No prep needed.
- Concrete: Begin with a layer of dry leaves and sticks around 3 to 4 inches.
- Grass: Add a layer of pre-soaked non-glossy cardboard, making sure each edge is well overlapped.
- Add a layer of newspaper overlaying each sheet until it is about half an inch thick; water the newspaper.

2. The second layer is alfalfa or straw, again between 3 to 4 inches and watered. Alfalfa is preferred because it is rich in carbon and nitrogen, but straw works just fine if that's all that's available.

3. A thinner, 2-inch layer of manure is added on top of the alfalfa, and you can add some compost, too, if you want. Water this layer.

4. Add 3 to 4 inches of straw and water.

5. Repeat steps 3 and 4, making the straw layer around 4 inches.

6. Make pockets in the straw, fill with compost, and add your plants.

IMPROVING THE QUALITY OF YOUR SOIL

Similar to creating soil, there are various ways you can improve the quality of your soil. Your choice may depend on the size of your garden and what you have available. Of course, the one thing that will remain consistent is no tilling! We will begin with how ecological succession works.

Not all soil is the same, regardless of soil quality. For example, the soil that you find in the forest will be dominated by fungi because of the surface decomposition. In contrast, bacteria will dominate the soil where weeds and perennial weeds grow. To improve the quality of your soil, you have to first understand the original conditions of your soil. There are three common conditions you might find:

Annual plants

Because they die at the end of their cycle, the plant decomposes and turns into mulch: food for bacteria and worms. To reproduce the perfect soil conditions, you should add compost and then maintain the soil with mulch like lawn cuttings, twigs, and straw. To boost diversity, you can rotate your crops.

Grasslands

Grasslands often have livestock on them, trampling the ground and adding manure which allows for a great balance between fungi and bacteria. Make sure that the soil is constantly covered with perennial cover crops. These will smother weeds and protect the soil underneath.

Food forests/permaculture gardens

If left untouched, ecosystems will end up like food forests (young mature forests). Plants die on the forest floor, and fungi and other organisms decompose them. An important relationship between plant roots and nutrients is developed with the fungi in the top few inches, allowing trees to get the necessary nutrients to grow. To boost this relationship, use green manures. You can also introduce mycorrhizal fungi to the soil by taking soil from areas with many mushroom-producing trees or scattering a dry spore mixture.

A very simple solution for taking care of your soil quality is to add garden paths and raised beds. Each time we walk on our soil, we compact it, making it harder for the air to circulate, not to mention the possible damage we do to the natural environment in the soil. Garden paths and raised beds protect soil from unwanted traffic on the soil.

Aerated compost tea

Before looking into aerated compost tea, it's important to know the difference between worm bin leachate and aerated compost tea. Worm bin leachate is the liquid that drains from the bottom of a worm bin. Leachate liquid is anaerobic, whereas, with aerated compost tea, we want a lovely rich aerobic liquid. If you want to use your worm leachate, make sure it is well diluted at a ratio of at least 1:10.

By steeping compost in water, we get a natural liquid fertilizer full of amazing beneficial microbes, including bacteria, fungi, protozoa, and nematodes. By delivering this nutritious boost in the form of a liquid, the plants can absorb the benefits straight up through their roots. When you add air to compost tea, you multiply the benefits by thousands. That being said, brewing compost only strengthens the nutrients and microbes already in the soil, so you must ensure that your original compost is of top quality. This means that it must be made from the correct compost materials and be well-aged. More advantages to aerated compost tea are:

- There is less need to use other types of fertilizer because the soil is absorbing more nutrients from the tea.
- More organic matter is broken down because there are more beneficial microbes in the compost tea, creating more diversity in terms of nutrients.
- Compost tea allows for a healthy soil food web, which can help protect both the plants and soil from pollution.
- There is greater moisture retention in the soil and, therefore, a more evenly moist soil. You don't have to water as often, and the plants are less stressed.
- Plants fed on aerated compost tea can have better immunity and be more resistant to diseases.

To make your aerated compost tea, you will need compost or vermicompost, a brewing vessel, a tea bag, and a pump.

The compost

As we said before, the compost must be of high quality. The greater the variety of organic matter, the more nutrients you will get, so make sure you are adding both carbon-rich and nitrogen-rich matter. You can use your vermicompost, and the good news is that because of the air, the worms will survive. It's still best to remove as many as you can, though. Don't use fresh animal manure or anaerobic compost because the smell will be pretty vile.

The brewing vessel

A simple plastic bucket will do, and the size will depend on your needs. Larger spaces might need tanks rather than buckets. Clean or new five-gallon paint buckets are ideal because they have the height to suspend the tea bag.

The tea bag

The tea bag will be a sack that holds the compost. It has to be breathable so that the microbes and nutrients can pass between the water and the compost. But don't use anything that is too large and will let particles pass through. Burlap, nylon paint strainers, old socks, or nylon tights all work well as tea bags.

The pump

The pump is what will provide air to the tea. For our plastic bucket, an aquarium tank pump is enough. It's even better if you have a pump with multiple ports in case you want to brew more than one bucket at a time. The tubes will then be connected to an air stone or bubbler inside the vessel.

- Fill your bucket to the top with dechlorinated water, leaving a few inches of space for bubbles.
- Fill your tea bags with compost (some people like to add kelp meal, fish fertilizer, biochar, or other nutrient

boots at this point) and tie them with a piece of string but leave the string long enough to hang the bag later.

- Work on the theory that two to five cups of compost is good for a five-gallon bucket.
- Dunk your compost bag in the water a few times to allow moisture to pass through the bag and agitate the contents. At this point, your compost will start to infuse.
- Use the extra length of string to tie the bag to the bucket's handle. You don't want the tea bag to be on the floor of the bucket.
- Add your air stone or snake bubbler to the bucket under the tea bag.
- Leave the air pump on for twelve to forty-eight hours and use the compost tea as soon as your brewing time is up.

If you want to empty the bags into the water, you can water your garden, and stir the contents well, so you don't end up with compost at the bottom of the bucket. If you are going to use a watering can or spray the garden with your compost tea, you can't mix the compost in because it may clog the sprayer. Use a fine-mesh strainer to prevent any particles from clogging your sprayer. It's best to use the tea after watering your garden. It sounds strange, but moist soil absorbs more than dry soil, and you don't want your compost tea wasted as runoff.

HOT COMPOSTING IN 18 DAYS

The composting method we have seen is cold composting and will take anywhere from six to twelve months to be ready for use. Now, I have already pushed your patience quite a bit, so let's look at how we can hot compost to reduce this time down to eighteen days.

Hot composting has other advantages. The compost becomes much finer, and you won't be able to distinguish between any of the matter that you introduce. Because of this fast aerobic process, you get the same volume of compost as the matter you introduce. Hot composting, or the Berkeley method, requires a little more attention to detail, and there are some specifics that you need to keep in mind.

- The compost heap needs to be about 3 square feet and 5 feet in height.
- The temperature of the compost needs to remain at 55–65°C or 131–149°F.
- The carbon:nitrogen balance needs to be 25/30:1 (remember the browns vs. greens).
- Contents should contain an even ratio of browns, greens, and manure.
- Large, high-carbon matter like tree branches need to be broken up first.
- Systematic turning is essential for thorough mixing; mix to turn the compost inside out.

You would be amazed at what you can hot compost. I have even seen people adding old leather to their hot compost pile. Hot composting relies on oxygen flowing through the matter. Once you have added all of your matter, water the pile well; you should notice drips of water at the bottom. Leave the pile unturned for four days. For the last fourteen days, you need to turn the compost heap inside-out every other day.

Here are some brown, green, and manure carbon:nitrogen ratios to give you an idea of the right balance.

Browns	Greens	Manuers
Wood chips- 400:1	grass/weeds- 25:1	Fish- 7-1
Straw- 75:1	Vegetable scraps- 25:1	Chicken- 12-1
Shredded paper- 175:1	Fruit waste- 35:1	Rabbit- 8-1
Dried leaves- 60:1	Coffee grounds- 20:1	Cow- 18-1
Fruit waste- 35:1	Urine- 1-1	Horse- 25-1

It's normal that this starts to get a little confusing, so you can use a general rule of adding one bucket of greens to two buckets of browns. This is also a good rule because it will help to keep layers of brown and green matter for the first four days without turning. For this example, manures are green.

Remember to always measure the core temperature in the middle of the pile. If it is too hot, smells, or reduces in size, you have too much nitrogen. Alcohol is a byproduct of anaerobic fermentation. If your compost is too hot and not turned for aeration, it can explode. It is imperative to monitor core temperatures and keep the compost pile aerated. When compost temperatures have stabilized, and contents have turned into a rich dark brown color with an earthy odor, the compost is ready to use.

MULCHING FOR BIG BENEFITS

Yes, mulching is more work, but the benefits outweigh the effort. What's more, the correct mulch will save you work in the long run. Remember the permie theory of bare soil being bad soil? Mulch removes this issue by adding a layer of organic materials to the topsoil that mimics a forest floor.

Aside from looking nicer than bare soil, mulch does an amazing job of retaining moisture and preventing erosion. It acts as a fertilizer and will suppress weeds, which can save time later on. There are different types of mulch.

- **Living mulch**: Living plants under the main crop. Annual plants like sweet alyssum and nasturtium work well under vegetables. Perennials like comfrey, thyme, and oregano make good living mulch for fruit trees. Living mulches are often referred to as cover crops, which we will look at further in Chapter 11 when choosing our plants.
- **Green mulch**: Weeds and other plants get cut at the base while the roots stay in the ground. The leaves get chopped into smaller pieces to fertilize the main crops. Green mulches can include comfrey, dandelion, chives, chickweed, and rhubarb leaves. Green mulch is also known as the grow, chop, and drop method and is a favorite because it mimics nature.
- **Leaf mulch**: Leaf mulch looks natural and is ideal for retaining moisture. Leaves can be collected and shredded, and left in a wire bin to use during the year. Don't add walnut leaves because they contain a chemical called juglone that can be toxic to most other plants.
- **Wood chips**: Wood chips look and smell lovely, and if you are lucky, you might find them for free. Perennials love wood chips but don't let the chippings come in contact with their stems. You can also use them with your vegetable garden and fruit trees.

An ideal combination is to use green mulch and leaf mulch. This way, you get the combination of weed suppression and fertilization. Additionally, like the compost bin, layered is better.

If you struggle with severe winters, it's a good idea to replace the layer of green mulch with a thick layer of animal manure and then top it with leaf mulch or wood chip. The combination of manure topped with different layers of mulch will help protect your plants from freezing temperatures and provide tons of nutrients.

There is no need to panic after all this information. I know that when talking to a newbie permaculturist, it is often like talking to the parents of a newborn baby! We look at our creation and only focus on everything that we are doing wrong. Your soil might not be in perfect condition, and you've probably made some mistakes, but your soil will forgive you. You now have multiple ways to improve the soil quality, but you don't need to start with every idea straight away. It takes time, but in time, you will develop a kind of understanding of the needs of your soil and know which methods are suitable for you.

CHAPTER 9

BUILDING GARDENS IN YOUR ECOSYSTEM

I LOVE this part of planning because it's where you can see your creativity coming to life. Remember that we still aren't looking at what plants and flowers to introduce, although you may have an idea of what you want. We aren't quite at the color stage but getting more into the details of your garden. You'll want to use another overlay for your garden design because it will help you to check that your gardens will align with the rest of your designs (most specifically, the water system) and still follow the permaculture principles.

RAISED BED GARDENING

Raised garden beds are popular because of their versatility, and they align with the permaculture principles on no digging. The size of your raised beds will depend on your design. Most go for around 4 feet wide because that is a comfortable distance to reach across. I have seen them as narrow as 2 feet placed along paths, which is pretty and practical. Height is another question that depends on what you plant. If you have short root plants like onions, you want to think about a minimum depth of 12 inches. For plants with longer roots, such as tomatoes, you can go up to 18 to 24 inches. There are no rules for the size of your raised beds, only recommendations. You might want rectangle beds, L-shaped, U-shaped, or keyhole-shaped garden beds. One of my favorite shapes for garden beds is a keyhole shape because it goes right along with the principles of permaculture by making the most use of spaces and edges.

There are several ways to make a raised bed. You might have an already elevated area of the garden that you want to add material to in order to make the raised bed. You may only want the bed to be raised slightly or, if you suffer from back problems, an elevated raised bed allows you to care for your plants while standing. Wood is a popular choice for making your own. Before you go and buy any materials, look to see what materials you have at hand that can be recycled. These could be bricks, stones, or sheets of metal.

If you use wood or metal sheets, you will need to join them together in the corners with either coach bolts and a corner piece, steel brackets, or spiked metal brackets. For longer raised beds, you will need extension pieces to join the wood or metal together for the sides. Thin timber strips can be fastened straight into the ground with U-shaped clips for a very simple and quick raised bed. Any material that you choose to use has to be non-toxic. Be very careful not to use treated or painted wood or tires. You don't want toxic chemicals seeping into your soil. You should also make sure they will resist harsh weather and are relatively resistant to sun and rain.

In the previous chapter, we looked at creating raised beds using layers of organic material to create a no-dig garden bed, but there are other options. You might want to use 50 percent of your garden soil combined with a top layer of 50 percent homemade compost. And, of course, choose your favorite beneficial mulch to help keep the soil fertile and moist.

You and your plants will enjoy having raised beds. The advantage for you is that your raised garden beds are higher up and can be less strenuous on your body. Also, with smaller planting areas, it's easier to create different soil conditions depending on the needs of the plants you are growing.

HÜGELKULTUR BEDS

A Hügelkultur bed (German for hill culture) is another type of garden bed that allows you to recycle garden materials without cutting them as you would have to for compost. You can make Hügelkultur beds with larger logs and branches. These are all slow-composting materials that break down into very stable humus. Humus will improve soil fertility and water retention, and because it is a slow process, the soil benefits for a much longer time. Once you have covered the woody material with soil, you have a mound ready to act like a raised garden bed.

Like a raised bed, the size will depend on your design and the amount of material you wish to recycle. The average Hügelkultur bed is approximately 6 feet by 3 feet around the base and 3 feet high. If you aren't looking for a mound in your garden, you can dig a trench to lay the logs and branches. A Hügelkultur bed requires layering your materials. Here are instructions for a simple Hügelkultur bed.

- Start with a thin layer of manure or compost on the area the logs will go. Soil bacteria can draw nitrogen down from woody materials, and the compost will help prevent this. Plus, it will improve your carbon:nitrogen ratio.
- Lay your logs and branches on top of the compost to

the height of 12 to 24 inches You need to compact the pile as much as possible with a spade or by carefully standing on it.

- Add a layer of straw followed by a layer of green mulch and finally, another layer of compost.
- The final layer will be about 1 inch of soil, followed by your plants. Create small holes in the mound and fill them with compost. Plant seeds, seedlings, or plants and water them well.

What you plant in a Hügelkultur bed does matter if you are looking for low-maintenance. You can plant anything in this type of bed, but generally speaking, and depending on your climate, Hügelkultur beds will be drier than the ground. You may want to think about planting things that don't like to get their roots wet and suffer from root rot. You can try planting onions, tomatoes, eggplant, beans, peas, rosemary, Smoke Bush, and Butter Daisy.

PLANTING A HEDGEROW

Hedgerows have many benefits apart from their visual appeal. It can shelter the garden from wind, noise, pollution, and of course, peering neighbors. Because a hedgerow has various types of plants, you can bring more diversity to your ecosystem. Hedgerows might be just one row of plants or up to 40 feet wide when there is sufficient space, and the length too will vary. When designing your hedgerows, here are some typical ways they can be used:

- To define property lines, paths, or desire lines
- To separate different sections of the garden, like a play area from a pond
- To create animal paddocks

- To manage water flow

To build a hedgerow, we will begin with the foundational plantings. Foundational plants are the tallest and usually trees, but it's more likely to be dwarf trees or shrubs in smaller urban gardens. Make sure you plant the foundational plants closest to the property line. To ensure that your hedgerow is nice and compact, only give taller plants 75 percent of the recommended space. After establishing where we want the foundational plants, we work our way in and start adding smaller plants as fillers.

Our supporting plants will again start with the tallest, just a little in front of your foundational plants, and staggered between them. Finally, ensure that each space will be filled out with your smallest herbs and flowers.

Notice that we haven't gone into too much detail about the types of plants you should choose. Generally speaking, hedgerow plants will be perennials, and the supporting plants are either annuals or perennials. Chapter 11 will aim to inspire you to choose the right plants for hedgerows. To give you a quick idea, here are some plants you can plant in order of height:

- **Tall trees**: oak, alder, black locust
- **Evergreens**: holly bushes, mahonia, junipers
- **Understory edibles dwarf/semi-dwarf**: apple, cherry, plum
- **Understory edibles bushes**: rose, currant, blueberry
- **Understory flowering trees and bushes**: lilac, flowering dogwood, witch hazel
- **Flowering herbs**: dandelion, yarrow, red clover

There are a few other tips for your hedgerows. When planting perennials, do it on a cloudy day as too much sun might shock them. When choosing plants for your hedgerow, make sure to select some nitrogen fixers so that your soil remains healthy. It takes anywhere from one to four years for a hedgerow to establish. In this time, work on building the soil; remember that extra water will be needed during the dry seasons, and keep adding mulch to control your weeds and retain moisture.

CREATING HERB SPIRALS

I have come across some amazing herb spirals that take full advantage of small spaces. Typically, an herb spiral will be between 5 and 6.5 feet wide, and this size works well because it allows for a growing space inside the spiral of 12 inches. Of course, the material you choose will make a difference, and bricks will add to the diameter, whereas metal will keep it smaller.

Herb spirals are so convenient because you can create numerous micro-ecosystems in just one spiral. The soil at the top of the spiral will be drier than that at the bottom. You also have a design that allows for sun and shade so that you can plant a variety of herbs from different climates.

Image: Herb spiral made from rocks

We are going to go over how to make an herb spiral with bricks. I like using bricks because you can take advantage of the warmth they store from the sun. The first thing is to decide on your location. Three things to remember: first, herbs love the sun. If possible, choose a spot that gets around eight hours of sun each day. Next, remember that you want to keep your herb spiral as close to the house as possible for convenience. Finally, the ground has to be level before you begin.

Draw out the design you want for your spiral. You can use a stick or baking flour. Lay down the first layer of bricks. For each layer, you want to begin about two and a half brick lengths from the first brick. This way, it will gradually increase in height. If your spiral is 5 feet wide, your final layer will be six or seven layers of bricks. Once the structure is ready, you can fill the spiral with a mixture of half garden soil and half compost. Add a thin layer of mulch to help with moisture and protect the soil from the elements.

For a small area, you have six core sections that different herbs will thrive in. Before you plant anything, be sure to understand which side of your spiral gets the most sun and shade.

- **Sunny + dry**: rosemary, lavender, aloe vera, sage, thyme, oregano,
- **Shady + dry**: nasturtium, yarrow, parsley, sage, thyme
- **Sunny + moist**: coriander, sorrel, basil, chives, strawberries, spring onion
- **Shady + moist**: sorrel, chamomile, rocket, borage, chives
- **Sunny + wet**: basil, dill, lemon balm, lemongrass

- **Shady + wet**: horseradish, dill, tarragon, chervil, parsley

Considering there are between seventy-five to one hundred different types of herbs, this is just an idea of what you can create with an herb spiral. It's always wise to check the plant information to see what conditions the herbs like before planting. You should wait until your seeds have become healthy seedlings before transferring them to the herb spiral.

One last word of advice would be to avoid adding mint to the herb spiral. Mint is fragrant, easy to grow, and medicinal with multiple uses, but it does like to take over space, so your mint is probably best in a separate location or a pot.

BUILDING TRELLIS SUPPORTS

Trellis supports are another fantastic way to make the most of every inch of your space. Not everyone has the space for fruit trees, but when we train them to grow in a particular direction, a fruit tree or any other tree can be grown in a narrow space. When training trees to grow along a trellis, it is called *espalier*.

Image: Espalier apple tree

The average espalier is 6 to 8 feet tall, so bear this in mind when deciding where to build the structure. If you are planting smaller trees, you can use a fence, but we like using a wall because of the heat from the bricks. If you are looking around your garden and you don't have a wall, don't worry, you can use the wall of your house. The only limitation is that your wall must have the full sun at midday—south-facing in the Northern Hemisphere and north-facing in the Southern Hemisphere.

There is no one rule for making a trellis, and there are plenty of different designs; you probably have materials in your garden that you can recycle. One easy DIY design would be to use wooden posts with wire mesh attached to them.

Another easy espalier support to make is by using two star pickets attached to a wall with right-angle brackets. Be sure to have at least 2 feet of the star picket in the ground for sufficient support. One end of the wire will be tied to the star picket, and the other end will be attached with a turnbuckle. Typically, you will make five rows of wire. Once all of your wires are attached, tighten them up, making sure that your structure is very sturdy.

As the tree grows, you need to train it to grow in the right direction. To do this, you can loosely tie branches to your wires and trim the branches, so they are encouraged to grow laterally, across the wire.

You can also use a trellis for your vines. Vines prefer a symmetrical structure to climb and cling to, so it is often better to use a wire mesh rather than rows of wire. You can either tie the vines to the wire or weave them through the holes. They might need to be tied in the early stages until you get some length to weave. Vines grow very quickly, so you might have to trim them more than once a year to keep them looking tidy and growing in the direction or design you want.

If you are anything like me, you will have your next overlay and be working out how to fit as many garden designs as possible into the site layout. This is great, but remember, Rome wasn't built in a day. It's always better to start small, watch what your garden tells you, and then add more.

CHAPTER 10

ECOLOGY AND WILDLIFE
HABITATS

IT WASN'T until I got the permaculture bug that I started to
realize how crucial all the little critters and other organisms are
to our ecosystems. We must understand and respect ecological
hierarchies within the world before introducing wildlife into our
gardens. Hopefully, in your observation stage, you will be able to
recognize the hierarchies.

- **Individuals**: An individual is one organism. You are
  an individual, so is your dog or cat, and the pear tree
  that is growing is also an individual.
- **Species**: Species are the type of individual. If we look at
  our individuals, we have humans, felines, and canines.
- **Populations**: A group of the same species living in
  your area. You might already have a population of bees.
- **Communities**: The populations of different species
  living in one area. Our goal is to create a community
  within our permaculture garden.

So what's the difference between a community and an ecosystem? An ecosystem includes the community and the environ-

ment, including abiotic (non-living) elements and biotic (living) elements. And why is this relevant? Because what we introduce has to work with the other species that are in the community. Just because you love beavers, you can't add one to your ecosystem without trees and water for the beaver to do what is natural to it—build a dam!

This chapter is where we will introduce ecological design. As humans, we play a role within the community, and our actions impact the ecosystem. We want to introduce as much biodiversity as possible but what we introduce should ideally be native to our area to encourage the development of positive and beneficial relationships. To help us with this, there are three ecological principles of permaculture to keep in mind:

1. Each organism plays a role in the community.
2. Understand the benefits of annual and perennial crops to improve soil conditions.
3. Increase biodiversity with as many plants, insects, and small animals as possible.

HOW TO INCREASE BIODIVERSITY IN YOUR GARDEN

Now, let's dive straight into how you can start introducing more crops and wildlife by having different types of gardens within your space. Not all of these spaces will be part of your design, and you don't have to feel guilty if you only want to introduce one or two. It's about doing the best you can with the space you have.

- **Kitchen gardens**: Window boxes for herbs, herb spirals, raised vegetable beds—things you keep in or close to Zone 1 for quick access.

- **Orchards**: Fruit trees don't need a lot of attention, so you may not think about creating them close to your home. However, remember if you have a high yield, it's a lot of work to bring the fruit back.
- **Pollinator gardens**: Approximately one-third of the food we eat relies on pollinators. We need insects like bees as much as our gardens do, so plan to plant flowers and shrubs that attract pollinators.
- **Field crops**: Normally part of larger designs, field crop gardens contain beans and grains.
- **Wild areas**: Typically Zone 5 of a permaculture garden. These areas are essential to understanding what is native to your ecosystem.
- **Zen gardens**: One function of your garden is to provide you with a space to relax. You might be happy having your wild area as your Zen garden.
- **Play garden**: If you have children, they need some space so they aren't tempted to run across your beds.
- **Accessible garden**: Again, it will depend on who is visiting your garden. If you are hoping to encourage community members to your garden, there has to be easy access for everyone, including the elderly and handicapped.
- **Hedgerows**: Though not a garden type, they encourage incredible biodiversity as well as multi-functions.

With these garden types in mind, we can look at how wildlife can be incorporated into the design.

WHERE TO START WITH YOUR WILDLIFE

As Graham Bell, the Chair of Permaculture Scotland, said, "Abundant wildlife is the first sign of a healthy planet." We want our permaculture garden to be healthy, therefore, wildlife is necessary. The more humans keep building, the less space there is for wildlife. What better way to incorporate the permaculture principles than to turn your garden into a sanctuary for animals.

Your observations on wildlife will have given you an idea of what is native in your area, but if your garden doesn't have the elements animals need to survive, you won't have observed them. Back up your observations with a little research on what animals thrive in your area, including types of birds, butterflies, rabbits, and squirrels. Depending on your area, you may have deer and even wild ponies. Next, you can check that you can provide all that your native animals need:

- Food supply, including flowers, insects, seeds.
- Shelter for themselves and their offspring. Leafy bushes and trees, bird/bat houses, and lizard shelters are some ideas.
- Fresh water in ponds, rain gardens, birdbaths, or even a bowl that can be topped off with rainwater by hand.
- Nesting areas. If they have food, water, and shelter, why would they want to leave? Provide quiet, private nesting spaces and add elements that your native animals appreciate.

BIRDS AND BEES FOR BIODIVERSITY

We already know that bees are essential for feeding the world's population and crucial for pollination, but birds are also excellent pollinators and they help spread seeds. When you consider the input and output for birds, we are onto a winner. Your input might be a seed holder, birdhouse, and birdbath. In return, you get pollinators, pest control, reduced weeds, free cleanup, and they can provide you with an early warning of weather changes and environmental hazards. There are an estimated 9,000 to 10,000 bird species and a whopping 25,000 recorded species. With the right environment, creating diversity with birds and bees is not challenging. We will look at how to encourage birds and bees in the following section.

The one thing we need to be cautious of with wildlife is disrupting the garden or overmanaging it. Everything in your garden has a reason to be there. You might be tempted to pull up the dandelion, but that is a favorite for bees. When aphids suck on a plant, they produce honeydew, a food source for wasps, bees, and ants. Why would you want wasps? They eat fly larvae! Lawn maintenance is necessary, and in some parts of the country, it is a code! But, mowing the lawn causes a lot of disturbance, whereas grazers do the same job naturally.

Wildlife doesn't have an address, and they aren't confined by a hedge or fence you put up. You will attract far more biodiversity if your neighbors are on the same wavelength. It will be hard for them not to be enthusiastic about doing their part to save the planet, but they might fear the work it involves. Offering to help them out gives you a little extra space and the chance to work toward a collective landscape.

How to Attract Pollinators

The first thing we want to avoid is monocultures. One type of flower or one color will only attract a small number of birds and bees. Perennials are best so that you have bright colors all year round. Different plants produce different types of nectar and protein—the principal diet of our pollinators. As well as a variety of colors, look for plants with different fragrances too. Try to incorporate as many types of plants as you can. If you want to attract native pollinators, the best way to go is to plant native plants.

The shape of flowers can also attract a diverse range of pollinators. Bees love flowers with ultraviolet markings called nectar guides, and I like to imagine these as landing strips that take them straight to the pollen, for example, primroses and pansies. Butterflies prefer flat-topped clustered flowers like white aster, and hummingbirds are more attracted to the funnel family of flowers such as salvia.

Trees, shrubs, and vertical structures also play their role. Wasps use resin and sap produced by trees to seal their cells. Shrubs and leafy plants are good for shelter. Just as stacking makes the most of vertical space for plants, it also creates different spaces for pollinators to co-exist.

A NEW APPRECIATION FOR INSECTS

Let's take a quick look at some common insects and how they can benefit our garden.

- **Ladybugs**: A fantastic pollinator that eats aphids, whiteflies, mites, and other plant-destroying insects. Plus, they are probably the cutest of insects.
- **Praying mantises**: The diet of a praying mantis includes grubs, aphids, grasshoppers, crickets, and flies.
- **Green lacewings**: They eat the larvae and adults of aphids, mealybugs, mites, and small caterpillars.
- **Dragonflies**: Control flying insects like mosquitoes, flies, and fruit flies and are attractive to look at.
- **Beetles**: They are good decomposers and like to feed on soft-bodied insects and insect eggs.
- **Spiders**: They help protect flower beds by controlling moths, wasps, beetles, aphids, and mosquito populations.

Like birds and bees, insects thrive on a variety of native plants. Trees with rough, peeling bark are the ideal home for insects because of all the cracks and shelters to hide in. Alternatively, you can make insect hotels.

An insect hotel can be a small wooden area that has sections that attract different insects. You can have one section for sticks, another for smaller logs, leaves, stones, pine cones, toilet rolls, bark, and so on. As you can imagine, the more materials your insect hotel has, the greater the diversity among your guests. Don't worry about the look of your insect hotel because insects aren't picky when it comes to aesthetics. Making an insect hotel is a fun project that the kids can get involved with too.

Image: DIY bug hotel project

If you are worried about waking up one morning and being overrun with insects, don't panic. A little further on, we will look at how to manage pests to maintain the ideal balance.

SMALL ANIMALS TO BRING YOUR GARDEN TO LIFE

Plants need animals to survive, not just for pollination but also for soil quality. By adding animals to the garden, we further increase the diversity in our garden and encourage the elements to flourish. Humans need animals for eggs, meat, milk, feathers, skin, wool, etc. So far, we have looked at the benefits of worms, chickens, birds, bees, and insects. You have to remember that having animals in your permaculture garden is not just a case of having pets that need extra attention. Animals love to work, and each animal you introduce must be done so with purpose and balance.

- **Snails**: Snails generally prefer dead garden matter, so they are good for cleaning up. Some snails will eat the eggs of garden pests.
- **Fish**: It's lovely to sit and watch fish, but they will also help keep your water balanced by eating plants and insects around the water. Their manure makes the water richer in nutrients that will benefit your soil if you use it for watering.
- **Ducks**: Similar output to chickens, but ducks tend to cause less damage to vegetable gardens, and they are excellent at controlling snail populations.
- **Rabbits**: Rabbits help to keep the lawn down and will happily eat weeds. Because their diet is mainly green, their manure is a rich fertilizer.
- **Geese**: Geese are great for food forests because they steer clear of vegetables, and their manure doesn't smell as bad as other poultry. Plus, they make a heck of a noise if an intruder comes.
- **Cats**: You won't need to worry about mice and rats if you have cats. They will also help keep the spider population under control, and while they don't often

eat cockroaches, they can kill them. Recent studies have found that stroking a cat and a cat purring can reduce stress and even lower the risk of heart disease.

- **Dogs**: The main reason to have dogs is for protection and security on your property. Of course, we have all seen how efficient dogs are at herding sheep, but they can be trained to herd and protect other animals. There are some impressive videos on YouTube of dogs herding chickens.
- **Goats**: Goats munch on weeds and shrubs, and when left in a small area, they will happily clear anything, including trees if you aren't careful. Goats produce milk, which can be used for making some delicious goat cheese.
- **Sheep**: Sheep are like walking lawnmowers that produce manure and they aren't quite as mischievous as goats. They also produce milk and wool.
- **Pigs**: An incredibly non-fussy eater, pigs will eat pretty much anything. They are also excellent diggers. If you put acorns or other types of nuts in the ground, they will forage for the acorns and dig out the deep roots of weeds.
- **Donkeys**: Naturally, the larger the animal, the more grass they will eat, and the more manure you can get. Their large manure is easy to collect, so it's simple to add it to your compost pile. Donkeys are hard workers and can help you to move materials or large harvests of crops.
- **Cows**: Cows are generally reserved for larger areas and it's recommended that you have at least one acre per cow. If you are lucky enough to have the space, you can get plenty of manure as well as milk that can be turned into cheese and other dairy delights!

Apart from food, water, and shelter, the animals you introduce need the appropriate amount of space, love, and attention. Finally, never forget the permaculture ethics that we extended to include animal care.

CONTROLLING PESTS

It's pretty safe to say that where there are plants, there will be pests. For millions of years before the first humans, plants and pests lived together in harmony without the use of any chemicals. Nature took care of the balance, something that chemicals significantly disrupt. When you use chemicals to control pests, you aren't only killing the pests but also predator insects that help to keep pests under control. As chemical pest control is only a short-term solution, without predator insects, the pests will return and often in larger numbers. As well as predator species, toxic chemicals might spread to neighboring crops, causing harm to other animals and people. Because pests develop a resistance to chemical controls, users have to rotate the products they use to be effective.

In permaculture, we turn to integrated pest management (IPM) systems. IPM is a systematic strategy that prevents, monitors, and controls pests with no pesticide use. Unlike chemical pest control, IPM looks at long-term prevention by understanding the environmental factors and managing the ecosystem. Monitoring your area will give you the chance to correctly identify the pests you have so that you can decide on the best solution. Finally, we can control the pests with one of the four IPM pest control types or a combination. The four categories are:

- **Biological controls**: Attracting birds and natural predators. Some predator insects include ladybugs, lacewings, parasitic wasps, and predatory mites.

- **Organic pesticides**: Permaculturists wouldn't use chemical pesticides even when it's the least toxic organic pesticides. We will always choose biological controls over pesticides. Some acceptable non-chemical organic pesticides include hot pepper, onion, and garlic sprays diluted in water with a tiny amount of soap to help with adhesion to the plant. These types of sprays should be used sparingly and only on small areas of the garden. If the pests persist, remember to focus on what nature is telling us rather than going against nature to fix what's out of balance.
- **Mechanical/physical controls**: Traps and cages are mechanical controls, and physical controls can be manually removing the problem, creating barriers, sticky traps, boiling water, or soil steam sterilization for diseases.
- **Cultural controls**: By changing the environment, we can improve the health of plants and the soil to control pests and diseases. For example, interplanting plants that repel pests and attract beneficial insects, planting trap plants, removing infected parts of plants, and taking away debris where pests like to set up home. Common plants that repel pests include Mexican marigold, chives, lavender, fennel, peppermint, and dill.

Apart from monitoring, preventing, and controlling, there is a more ecological approach to IPM. Once you have identified the pest as best you can (take a look at the type of damage they cause because this will help), you can find out as much as you can about its preferred environment, life cycle, and what it dislikes. You can then make the necessary changes to your garden so that the pest migrates to more favorable conditions. It could be as simple as taking away their food, water, or shelter. If this doesn't work, you can then look at the best control method.

Now, IPM will still take up your time, and the goal of permaculture is to have a garden that's mostly self-automated. This is why we keep going back to the importance of choosing natives. Native plants and crops are perfect for your soil and climate, attracting the best insects to control pests naturally. The more diversity you plant, the more natural pest controllers will be attracted to your area. Many of the small animals we looked at in the previous section can be natural pest controllers too. Try to incorporate fruits and vegetables that are harvested at different times of the year. There are two benefits to this, you can produce food all year round, and you don't risk having all your ripe fruit and vegetables ruined by one insect infestation.

Don't worry if you get some negative feedback. It might be the case that you have introduced too much, too soon, or not enough. Pay close attention to any wildlife you introduce; take notes so that it's easier to remember and make gradual changes. Now that you know what animals you want, it's time to look at the different inputs we can provide for them and ourselves as well as those final steps to creating a beautiful permaculture garden.

CHAPTER 11 (PART 1): PREPARING THE PLANT GUILDS

WE HAVE REACHED the stage where we add the icing to the cake, the details, the color, the life into our permaculture garden. Try not to get carried away. It's so easy to pop down to the garden center and load up a trolley with all the most amazing plants but remember that everything has to have multiple functions. It's worth mentioning at this point that many people will use the terms guild and food forest interchangeably. The principles are the same. The main difference is that a food forest generally includes the entire garden and a guild primarily includes a specific set of companion plants.

When you decide on your plants, the first thing to do is make sure they align with the permaculture principles. To narrow the choices down further, think about the following questions:

- What crops do you like to eat?
- What plants are difficult for you to buy?
- What plants are expensive to buy?

The answers to these questions will guide you into choosing plants that provide benefits for all. The theme throughout the whole book has been polyculture over monoculture. Our ecosystems need as many different plants and flowers as possible to increase diversity and mimic nature.

It might seem easier to buy plants as growing from seeds doesn't guarantee an outcome. In many ways, it's easier to grow plants from seeds because the packet contains all the information you need regarding conditions. With that in mind, let's get straight into guilds.

WHY GUILDS ARE PERFECT FOR PERMACULTURE

Guilds tick all of our boxes. They require polyculture with companion plants that benefit each other while mimicking nature and are, therefore, low-maintenance. In your guilds, you should aim to have these seven main components:

1. **Food for the family**: There are six food groups, including fruits, veggies, staples, legumes and nuts, fats and oils, and animals. We can try to incorporate as many food groups as possible to increase diversity and your health benefits. You may also want to plant things for medicinal purposes, which we will discuss later on.
2. **Food for the soil**: Feeding the soil will encourage healthy soil and a better yield rich in nutrients. Food for our soil includes compost, mulch, fertilizers from animals, teas, and nitrogen-fixers (more on this later).
3. **Diggers**: We need diggers to open up the soil and allow air and water to absorb better. Deep-rooted trees are one solution, as are root crops like carrots. Diverse deep roots mean the soil is dug at different depths and widths.

4. **Groundcover**: Groundcover can be layers of mulch or crops. Remember, bare soil is bad soil, and groundcover helps retain moisture and protect the soil from the sun.

5. **Climbers**: Beans, cucumber, grapes, or passion fruit will provide food for the family and take advantage of vertical spaces.

6. **Supporters**: Climbers need support, but if you choose to use trees, bushes, or stalks as support, you need to make sure they are good companions or risk one taking over the other. Trellises are supports, as are fences, a birdbath, or your house.

7. **Protection**: Your guild might need protection from the elements, like strong winds. However, it is more likely that it will need protection in the form of pest control using integrated pest management, whether that's using flowers or other insects.

We can combine the seven components with the seven layers of a guild to maximize vertical space. The top layer is the canopy. On a larger scale, canopy trees could be trees used for lumber, and on a smaller scale, you can plant fruit and nut trees. These trees can also be used for the vertical layer as support for your climbers. The next layer is low trees such as dwarf fruit trees, followed by shrubs like currants, berries, and hard-stemmed shrubs (I like rosemary and oregano). After the harder shrubs, you have the herbaceous layer. You can plan for soft-stemmed bushy herbs, flowers, and vegetables. For the groundcover layer, there is a wide range of herbs, salad leaves, strawberries, etc. Finally, for your diggers, you can plant root vegetables.

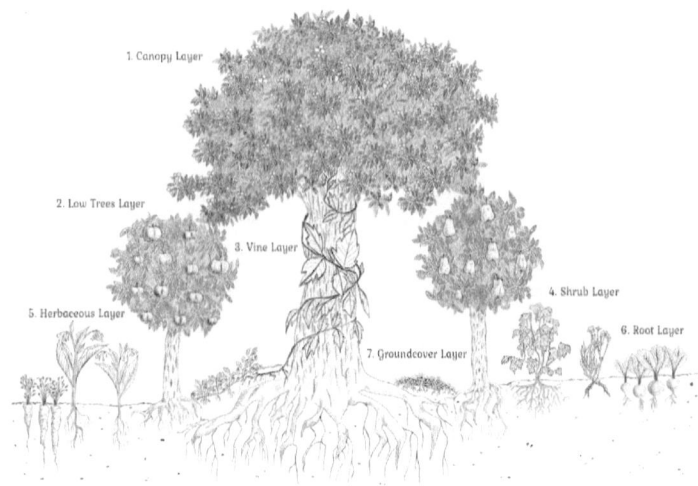

Image: 7 Layers of a guild

How to build a guild

At the center of your guild will be the anchor. Depending on the size of your guild, you might choose a vegetable plant or a fruit tree. Around the anchor plant, dig a ditch. The ditch will improve moisture and can be used as a sunken bed for more crops. You have to mulch the ditch well so that the roots of your anchor plant aren't exposed.

From this anchor, you will choose companion plants to go around it. Don't forget to choose a location that is best suited for the anchor plant in terms of soil, irrigation, and sun.

Everything you plant in your guild should solve any existing problems in the location, which you will have spotted during your observations. For example, if you have an area that has too much water, you would want to add thirsty plants like Monkey flower and Indian grass.

It's essential that you make use of all of your space. You don't need neat rows or clean edges. Because each element will be a companion plant, they want to be close together and any bare spaces filled with groundcover. You can even take advantage of spaces to plant the crops you find interesting, although you have never eaten them before, or an herb that you often see in recipes but had never thought to plant.

Guilds don't require a certain amount of space. You can create one as small as a window box if that is what is available. What is crucial is the pattern and the planning. Decide what is important for your needs, check that these elements have three functions, and then find the companions for these elements. Do these plants meet your needs? If so, find their companions and check that they have at least three functions. There are so many options out there that you will eventually come up with the perfect guild for your space.

16 BENEFICIAL PERENNIAL PLANTS FOR PERMACULTURE

Many people comment on permaculture being overwhelming, at least in the first few stages of getting started. I think, more often than not, it's more a case that we are spoiled for choice, especially if we have limited space. Out of all the possibilities, how can you narrow down what to plant? Observation, native plants, and companion plants will help us to become more selective, and I have always found that the following sixteen plants are always a great start.

1. **Comfrey**: Seen as a weed by some, it is a pretty plant with numerous benefits. Its roots can extend down 6 to 8 feet, bringing deep minerals to the surface to feed other plants. Extracts from the roots and leaves contain

allantoin, which boosts new skin cell growth for healing and has Rosmarinic acid to relieve pain. The leaves are also rich in nutrients and make for an excellent chop and drop mulch or compost accelerator.

2. **Hazelnut**: These bushes or small trees can handle the shade and act as good windbreakers. They have a long lifespan of up to fifty years and, in that time, can produce plenty of nuts that can be turned into oil or flour.

3. **Jerusalem artichoke**: Neither an artichoke nor from Jerusalem, they are part of the sunflower family and will grow tall. They are hardy and need little maintenance. You can eat the tubers which are high in protein, and use the stems as support for your climbers. Take care where you plant them, as they can spread quite aggressively.

4. **Fiddleheads**: This fern can be cooked. They are rich in vitamins and you can take advantage of them in spring before other crops are ready to harvest. Fiddleheads like wet soil and shade.

5. **Mulberry**: A messy and delicate berry that is a good source of food for your wildlife—and you can make a great jam from them. They have large leaves for shade and strong branches, so as well as providing food, the mulberry tree makes a good home for wildlife.

6. **Arrowhead**; An aquatic plant that provides shade around a pond and tubers that taste earthy and nutty. The tubers cook like potatoes and can be quite versatile.

7. **Mint**: From teas to sauces, helping indigestion, and potpourri for your home, there are plenty of uses for mint. It will attract bees but the smell repels many other pesky insects, and mint does spread, so it makes a good living mulch.

8. **Red clover**: Three simple functions of red clover are food for livestock, a cover crop, and a nitrogen-fixer.

9. **Wild leeks**: Another spring harvester, wild leeks are a good source of food that can be pickled and preserved. Wild leeks grow best under deciduous trees. They need shade, especially in summer, because they don't like heat.

10. **Asparagus**: Incredibly low-maintenance but does take two to three years before you can harvest the crop. Asparagus and tomatoes are best friends. Asparagus repels nematodes, a parasite harmful to tomatoes and tomatoes repel asparagus beetles.

11. **Edible flowers**: They attract pollinators but may also repel rodents. Use flowers like hibiscus, magnolia, nasturtium, and violets to dress your dishes like a professional.

12. **Stinging nettles**: Rich in nutrients and a favorite for medicinal uses. Once you boil stinging nettles, the sting is removed but the high levels of vitamin C remain. This plant has been used for pain relief, eczema, anemia, and much more.

13. **Strawberries**: Unlike mulberries, strawberries are less messy, and unlike raspberries, there are fewer brambles. Being a groundcover, strawberries also make a good living mulch.

14. **Walking onions**: Many onions are annuals, but the walking onion is a perennial. As the bulb grows on top of the plant, the weight pulls the bulb down, and it replants itself.

15. **Black locust**: Another excellent example of the stacking functions, the black locust is a strong tree that can be used for erosion control, a windbreaker or durable timber. Honeybees love the flowers and the seed pods are nitrogen-fixers.

16. **Hops**: If you are interested in brewing your own beer, you should try this climber. You can also dry the hops and add them to lavender to make your home smell nice. Hop vines are fast growers and give other plants lots of shade, and the flowers will attract pollinators.

As you list the plants you want, try to incorporate as many low-maintenance plants as possible. Some of the plants need to be deep-rooted for the soil, and others need to be nitrogen-fixers. There should also be plants with a lot of foliage for creating mulch. With this combination, you know that your soil will maintain its health.

FRUIT TREE GUILDS

I am a big fan of fruit tree guilds and their importance to permaculture design. I encourage people to create a fruit tree guild regardless of their garden size because you can create a mini-ecosystem just within the guild. While talking about size, as the anchor of your guild will be the fruit tree, it's handy to know the average size these trees can grow to so that you can plan the optimal size for your garden.

- **Ultra-dwarf fruit trees**: The Dorsett golden apple is an example of an ultra-dwarf fruit tree that grows to around 3 to 6 feet.
- **Dwarf fruit trees**: Trees like apricot, peach, and nectarine grow to an average of 6 to 8 feet. There are some taller dwarf trees like sweet cherries that grow up to 14 feet.
- **Semi-dwarf fruit trees**: These are a good size because you get a larger yield, but they are still a manageable size. Trees like the Kieffer pear tree grow to between 14 and 22 feet.

- **Standard fruit trees**: There is a good range of sizes. A standard peach tree can grow up to 15 feet, apricots up to 20 feet, and apples up to 30 feet.
- **Citrus trees**: Citrus trees can be dwarf to standard in size. If you create a fruit guild in a container, the dwarf citrus fruit can grow to just 6 feet, whereas standard-sized citrus trees can be as tall as 18 feet.

One of the most popular fruit tree guilds is the apple guild. After planting your apple tree as the anchor, you can plant some of the following to complete the guild, ticking the boxes of various layers and all with a minimum of three functions.

- **Shrub**: blackcurrants, oregano, raspberries
- **Herbaceous**: dandelions, marigolds, onions
- **Diggers/Rhizome**: carrots, radish, beetroot
- **Groundcover**: clover, nasturtium, parsley
- **Climbers/Supports**: sunflowers, cucumber, roses

To plant a fruit tree, you need to first decide on your location. Find a spot that receives the most amount of sunlight. The more sunlight your tree gets, the more energy it has to produce healthy fruit. Never plant a fruit tree in summer as the roots won't have grown enough to access moisture that is deeper in the ground. A deciduous tree can be planted in winter and, if it is actively growing, in spring or autumn. The ideal time to plant an evergreen is in the spring, and second to that is autumn. If you are planning to mix evergreens and deciduous trees, be sure to plant the deciduous tree closer to the sun as it needs warmer soil to bring it out of dormancy in spring.

As you can imagine, soil preparation is crucial for planting trees. You would never build a house without foundations, so take time to get your tree's foundations right. It doesn't take a lot of time, and all you need is a spade/trowel, compost and aged manure, and stakes and tie if the tree needs extra support.

- Your tree will come in a pot, the area inside the pot is what we call the root ball. Begin by digging a hole the same depth as the root ball and three times the width. Pop the pot in to make sure the soil in the pot is level with your soil.
- In the base of the hole, mix some of your compost into the soil.
- The soil you have dug out to make the hole can now be mixed with compost and aged manure. Use a ratio of 7 parts garden soil, 2 parts compost, and 1 part manure. When mixing the soil, you can also add kelp meal, worm castings, and biochar to give the tree and soil an extra boost.
- Place the tree in the hole, check that it is sitting upright, and double-check that the root ball is level with the soil.
- Fill the remaining space with your soil mixture, but you don't want to pack it in tightly because compacting the soil restricts the airflow. Water the newly planted tree. If you can, add compost tea that has been brewed with mycorrhizae, molasses, fish fertilizer, and worm castings for another boost to the soil microorganisms and tree vitality.
- If you feel that the fruit tree will benefit from some support, add two stakes, one on either side of the root ball area. If you go too close to the trunk, you risk damaging the roots. Use soft material to tie the trunk to the stakes.

Add September and March to your permaculture calendar as months to add some natural fertilizer to the fruit tree. You can use compost or worm tea to give the trees a burst of nutrients. Don't panic if you don't see fruit in the first year. No fruit isn't negative feedback; it's just that the fruit tree is using all its energy to establish new roots.

I have heard so many stories of pruning fruit trees that conjure up the image of Edward Scissorhands. Pruning is necessary, and it has to be done in the right way. You might think that dwarf trees are the best for smaller spaces, but they also require proper pruning to maintain their size. If not pruned, some fruit trees can grow as tall as a standard tree. How you prune your fruit tree in the younger years will teach it how to establish and grow. Keeping the tree well-maintained will make it easier to pick and produce a larger yield and a higher quality of fruit. Pruning fruit trees also allows for space between other elements so that air can circulate and light can pass through, and there is less risk of disease spreading.

If you are pruning branches that are thinner than your thumb (and this is normal for young fruit trees), you can use secateurs or a small hand saw. When you plant your tree, decide on the height you want the tree to be, and cut back the top and side branches by two-thirds. You need to locate the branch collar (the bulge under a branch where it attaches to the tree) and the branch bark ridge, like a tree's wrinkles above the branch. Cut to the outside of the branch ridge but not into the collar. This is where new cells grow and seal off the wound after pruning.

When spring comes, you need to prune the new growth back by half. In summer, there will be new growth that can be again cut back by half. In the second year, pruning will be the same. If you have a more vigorous fruit tree, you can prune in spring, early summer, and again in late summer.

In the third year, we prune the tree back to the height we want, and this might involve completely cutting back some branches.

If there are any dead, broken, or diseased branches, completely remove those too. Fruit-bearing branches need to have 6 inches of free space around them, so prune to create this space they need. Some branches may need to be removed to ensure others can grow properly. For example, if you have two branches growing together in the same direction or when one branch grows across another.

The time and effort you put into these first three years will set you up for a well-established fruit tree for years. What's more, as your fruit tree ages, it will produce more fruit. For fast results, you have your herbs and vegetables, but you need a healthy dose of permie patience for fruit.

BE WISE ABOUT YOUR WEEDS

 Weeds are flowers too, once you get to know them."

A. A. Milne

We have long been told that there is no place for weeds in our garden. Yet weeds have been around for a lot longer than we have and deserve a spot in the permaculture garden, particularly those that provide at least three functions. Here are just some types of weeds that will bring far more benefits to your garden than harm.

- **Bashful mimosa**: All parts of this weed have been shown to have health benefits as they are rich in antioxidants. It can be used as a groundcover for tomatoes and peppers, and it attracts predatory beetles.
- **Caper spurge**: It can be used as a food source for your livestock and moles can't stand it. In French folk medicine, this weed is used to aid digestion.

- **Cow garlic**: A wild cousin of onions and garlic, it makes a great companion for fruit trees, cabbage, broccoli, and carrots. It can be used to keep slugs and aphids away, and you can use it in cooking like you would chives.
- **Wild mustard**: When used as a companion for your brassica crops it will trap pests. It can also be a good companion of grapevine, and you can eat the seeds and the leaves.
- **Wild rose**: Some have used wild rose as a natural fence for their livestock; it deters rodents and deer and, at the same time, traps Japanese beetles. Rose hips are full of vitamin C and can be used in herbal teas.
- **Borage**: Borage leaves add a nice crunch to your salads, and the flowers are also edible. Predatory wasps are attracted to the flowers, and herbalists commonly use this weed to cure a cough, fever, and even depression.
- **Dandelion**: Dandelions help all other plants because their taproots break up hard soil, and it's a favorite for pollinators and rabbits. Plus, all parts of the plant are edible.
- **White clover**: This is a nitrogen-fixing groundcover and this weed accumulates phosphorus, which is vital for all plants, especially seedlings and young plants. The flowers are also edible and attract ladybugs and other pollinators.
- **Lamb's quarters**: This is an excellent nutrient accumulator for nitrogen, phosphorus, calcium, and potassium. The deep roots will help provide nutrients to surrounding plants. The edible leaves are high in vitamins A and C, fiber, protein, omega-3, and omega-6 fatty acids.

This is a shortlist of potential weeds that might spring up. The main idea is that you research the type of weed before pulling it up. On a similar note, there are weeds that you don't want either because they are invasive, poisonous, or they are prone to disease.

Atropa belladonna (deadly nightshade) has beautiful white or purple bell-shaped flowers that cause hallucinations and can be deadly. Don't be fooled by its beauty, and wear gloves when touching it. Even the leaves can cause blisters. Bittersweet nightshade isn't as toxic but can still be deadly for pets and children. Giant hogweed is like a poisonous carrot for the Big Friendly Giant, growing as tall as 13 feet. Then there are the more well-known toxic weeds like poison oak and poison ivy. Purple loosestrife, Japanese Honeysuckle, and Japanese Knotweed are highly invasive, so if you choose not to remove them, you will need to take extra care with maintenance. Mallow, jimson weed, and tree tobacco (a plant that grows like a weed) are reservoirs for viruses and diseases, and the symptoms might not always be visible.

CREATING A MEDICINAL HERB GARDEN

I have always been drawn to growing particular herbs and plants for medicinal purposes. Aloe vera has antioxidant and antibacterial properties, and the gel can be used to improve your skin and digestion. It can relieve sunburn and heartburn and help your cuts and wounds heal. Nevertheless, what we grow for medicinal purposes won't have the same effect on everyone. If you have medical conditions, you should check with your doctor to ensure they don't react with other medications.

For boosting your immunity, especially during cold and flu season, you can try things like astragalus, echinacea, and elderberry. Astragalus is a low bush or stalk legume plant, and you can save quite a bit of money growing this. The roots can be added to soups and stews, and it is a nitrogen-fixer. Echinacea has lovely shrubby flowers that attract pollinators and can help relieve a cough, acne, and toothache. Elderberries can't be eaten raw, but when prepared, they may lower fever and relieve a cough. It's high in antioxidants, and you can use elderflower from the same bush for juice.

If you constantly feel on edge, you might want to plant lemon balm, which is part of the mint family. This plant is antiviral and a relaxing nervine. There have been studies to show lemon balm can reduce the symptoms of ADHD. Next is tulsi, a.k.a. holy basil. Tulsi is used to boost immunity, and the benefits of tulsi increase the more you take it. Aside from relaxing you, tulsi can help the body oxygenate itself and improve heart health and memory function. Passionflower is a stunning vine herb that produces passion fruit. The flower is used to help with relaxation, overcome anxiety, and ladies, it may also help with period pain. Don't eat the leaves or unripe fruit as they are toxic.

Thyme, sage, and garlic are antimicrobial herbs that can be added to lots of different dishes. Garlic is easy to grow; just pop a clove into healthy, fluffy soil and harvest when the leaves start to turn brown. Garlic can boost the immune system while lowering blood pressure and cholesterol. Sage smoke can be better than an air purifier, reducing pathogenic bacteria by 94 percent in a confined area and keeping it that way for thirty days (Nautiyal et al., 2007). Thyme (both wild and creeping) is an antibiotic and a disinfectant.

Dandelions are packed full of nutrients and can be made into a tea to help digestion. Ginger has to be a personal favorite because it has so many health benefits. Ginger is a rhizome that you can plant to fill a space in your guild. Ginger is well known as a reducer of dizziness and nausea, but it can also warm you up because it improves blood flow. You can use it to relieve pain from inflammation, and it may help prevent cancer. Ginger is delicious and adds a lovely kick to many dishes.

Hawthorn's fruit, leaves, flowers, and stems can be made into teas, vinegar, and jellies for a healthy heart. It may help with stress and fatigue too. Coriander is another easy-to-grow herb that can promote a healthier heart.

When deciding where to plant your herbs, you can also work in zones. Your culinary herbs are best kept closer to your kitchen in Zone 1. Zone 2 herbs are those that work well as companions for your vegetables. Medicinal herbs and mints are better in Zone 3 where there is more space.

Please remember, and I can't stress this enough, drinking hawthorn tea will help your heart, but it doesn't mean that your body doesn't still need exercise. Elderberries can help a cough, but if the cough persists, it can develop into a chest infection, and you may need antibiotics. The key words in this section are boost, increase, can, and may. Everyone is different!

THE NEED FOR NITROGEN FIXERS

Nitrogen is essential for photosynthesis, and it improves plant growth, especially for leafy greens, flower buds, and fruit growth. Alongside potassium and phosphorus, nitrogen is used in the highest of quantities. But the problem is, it's easy for plants to absorb all the available nitrogen, plus we lose more with water and sun. We have dipped in and out of nitrogen fixers with some examples, now let's look at the science behind this gold dust for gardens.

If you can remember back to those science classes at school, you might remember that most of the earth's atmosphere is made up of nitrogen, but this doesn't help plants because they can't take in nitrogen from the air. Plants take in nitrogen from the soil.

So, atmospheric nitrogen needs to be converted into nitrates in the soil. This process occurs when there is lightning and nitrogen-fixing bacteria, as two examples. Another example is when plants die and decompose, and nitrogen is returned to the soil. As you can imagine, the process is somewhat complex, and the most important takeaway is how we can make the most of nitrogen fixers to improve the quality of our soil and produce healthy crops.

You can buy tests that will indicate the nitrogen levels in your soil. There is also feedback that you watch out for when there is a nitrogen deficiency:

- Slow or stunted growth.
- Leaves that are smaller than average.
- Leaves closer to the ground turn yellow.
- Leaves higher up start to turn yellow over time.
- Flowers are smaller than average and die quicker.
- If you do get fruit, it will be smaller.

Animal manure—particularly cow, rabbit, and chicken—is high in nitrogen, so it's a good idea to add this to your compost. Coffee grounds can also be added to the compost or straight into the soil. One that might surprise you is hair. Human hair contains 15 percent nitrogen (*The American Journal of Clinical Nutrition*, 1967). As well as this, hair decomposes slowly, so it releases nitrogen over long periods of time. The next time you empty your hairbrush, push the hair into soil that seems to be suffering. We can also make sure that various parts of our guilds and gardens contain nitrogen-fixing plants.

Legumes are nature's best nitrogen fixers, so plant peanuts, cowpeas, soybeans, and fava beans when possible. These legumes have the highest amounts of nitrogen, but practically all legumes are nitrogen fixers. Lentils would be an excellent option as a food source too.

Let's take a look at three categories of nitrogen fixers that can be incorporated into different layers of your guild.

Nitrogen-fixing trees (options for your anchor or canopy):

- Alder
- Siberian pea tree
- Mimosa
- Redbud
- Kentucky coffee tree

Nitrogen-fixing shrubs:

- Autumn olive
- Russian olive
- American Bayberry
- Cliffrose
- Mountain mahogany

Nitrogen-fixing herbaceous plants"

- Beans and peas
- Licorice
- Alfalfa
- Wood vetch
- Clover

Just remember that each plant's ability as a nitrogen fixer will depend on the quality of your soil, as well as your region and climate—for example, not all legumes like the same growing conditions. Broad beans and lentils are cool-season legumes, while cowpea and the common bean are warm-season legumes. A plant will do its best work when it can thrive!

A Closer Look at Cover Crops

I have mentioned cover crops quite a bit thus far, and it's finally time we discuss planting and caring for cover crops. In the next chapter, we will go over companion plants, and I will include more ideas of what to grow with other elements of your permaculture garden.

Your cover crops have an abundance of benefits for the entire ecosystem. Their main job is to protect and enhance soil quality, but cover crops will also help with pest management and also provide food for your livestock and pollinators.

There are three main types of cover crops: root producers, nitrogen fixers, and biomass producers. Deep root producers such as diakon radish and turnips have large roots to break up the soil and improve water filtration. Nitrogen fixers are planted to help fix soil nitrogen levels and include legumes such as clover, alfalfa, peas and beans. Then we have the biomass producers like wheat, rye and buckwheat that germinate quickly, grow abundantly, and help with weed control.

The easiest way to plant cover crops is from seeds. You can spread the seeds as you would grass seeds, a technique referred to as throw sow. It helps to scratch the soil's surface first to encourage level soil and allow the seeds to settle better (it's like fluffing up your pillow). Throw sow won't work as well if you are planting root vegetables or beans as a cover crop because they need to be planted further apart.

The trick to cover crops is choosing the right time to plant them. They need time to grow before the first frost, but you also don't want them growing enough to produce seeds. Generally speaking, this is around four weeks before the first frost unless you are planting rye, which will handle any time.

It is essential that cover crops are cut before they seed. Exactly when will depend on your region and the crop you have planted, and if they overwinter or not. Plants that overwinter don't cope with the cold and need to be taken inside or covered so that the frost doesn't damage or kill them. Cover crops that are overwintered need to be cut back to ground level in early spring, whereas those that don't overwinter can be left to cut until after the first frost. In both cases, you need to cut at the base of the plant so the roots are left intact. You can leave the cuttings on the ground as mulch or add them to your compost pile. If it is coming up to winter, leave them on the ground to protect the soil from the cold weather.

At this point, you may feel like there will be a lot of work, and the more you plant, the more work there will be. The best way to reassure you is that it's a little bit like having two or three children closer together in age. It may be more difficult in the early years, but as they grow up, your children start to entertain and even take care of each other. By paying close attention to companion plants, you will find that the workload is reduced because each plant helps to care for the next.

Chapter 11 (Part 2): The Many Benefits of Companion Planting

Part 2 of Chapter 11 will be a bite-sized chapter with lists rather than information on the how and when. Also, remember that this is just an idea of some of the ideal companion crops. The Royal Botanic Gardens has approximately 390,900 plant species listed, and to list the companions for all of them would require a book on its own. In the same way, planting information on each combination would frankly make for a boring book. Use the information on seed packets for planting, and don't feel you have to plant something just because it's a good companion. What you plant has to meet the permaculture principles but also your needs. Search the Natural Capital Plant Database website (www.permacultureplantdata.com) to find detailed information and specifications for a specific plant.

What Goes with What?

- **Apricot**: You can use an apricot tree as an anchor and surround it with basil, tansy, marigolds, garlic, nasturtium, spinach, stinging nettles, and sunflowers.

They aren't good companions for tomatoes potatoes or yarrow.

- **Asparagus**: Good companion plants include basil, cilantro, marjoram, parsley, strawberries, and tomatoes. Best to keep separated from chives, onions, leeks and garlic.
- **Basil**: Basil grows well with apricots, asparagus, chives, cucumber, and fennel. Keep basil apart from rue and Swiss chard.
- **Beans**: Beans are great companions for many plants such as broccoli, brussels sprouts, cabbage, carrots, and cauliflower. You may also consider corn, cucumber, lettuce, grapevine, marjoram, parsley, peas, potatoes, rosemary, sage, savory, and zucchini. Keep your beans away from fennel, chives, garlic, and onion.
- **Broad beans**: The only difference between broad beans and beans is that they won't pair with grapevine so well or carrots.
- **Bush beans**: This is very similar to beans with a few differences. Bush beans pair well with beets, celery, and cucumber. You can also try spinach, strawberries, and sunflowers. Again, they aren't a fan of carrots.
- **Climbing bean**: Climbing beans are much like beans, but they aren't too keen on carrots. They do, however, like radish.
- **Beets**: Remember, when pairing with beans, stick to bush beans. Aside from this, they like broccoli, Brussels sprouts, cabbage, cauliflower, lettuce, marjoram, and onions. You can plant beets next to Swiss chard but not tomatoes.
- **Borage**: Plant borage with broccoli, Brussels sprouts, cabbage, cauliflower, cucumber, and peas. Other good friends will be squash, strawberries, tomatoes, and zucchini.

- **Cabbage**: As mentioned, cabbage gets along with all of the beans, beets, and borage. It is a great companion for chamomile, celery, coriander, cucumber, lettuce, marigolds, marjoram, and nasturtium. In fact, you can also plant it next to onions, peas, radish, rosemary, and sage. Keep cabbage away from garlic, rue, strawberries, and tomatoes.
- **Carrots**: Carrots tend to get along with a nice variety of garden elements, including beans, chives, coriander, cucumber, lettuce, marigolds, marjoram, onions, parsley, radish, rosemary, sage, and tomatoes. Carrots don't get along with dill
- **Celery**: Celery isn't good friends with potatoes. Apart from that, plant it next to bush beans, broccoli, cauliflower, kale, kohlrabi, leeks, onion, marjoram, mustard, peas, radish, turnip, and tomatoes.
- **Cherry**: Cherry trees are so pretty in blossom and go nicely with chives, lettuce, marigolds, nasturtium, spinach, and squash. Don't plant potatoes near cherry trees.
- **Chervil**: Also known as French parsley, chervil will go well in a spiral herb garden with coriander and parsley. You may also want to plant it with garlic, lettuce, radish, or yarrow.
- **Corn**: Corn is happy with many companions like amaranth, all beans, cucumber, dill, lettuce, geranium, peas, potatoes, pumpkin, squash, and radish. You shouldn't plant sage or tomatoes with your corn.
- **Eggplant**: Eggplant is a little fussier, and prefers to be with with fragrant herbs, beans, marjoram, marigold, potatoes, and tomatoes.
- **Fennel**: This sweet-smelling herb complements basil and thyme. You want to keep it away from your beans.
- **Grass**: Mulberry is a nice combination with grass, as

are tomatoes and yarrow. Take care not to grow cherry, other fruit trees, and sage with grass.

- **Horseradish**: Horseradish is a great companion plant for apples, apricots, cherry trees, potatoes, rosemary, stinging nettles, and yarrow.
- **Lavender**: This fragrant herb is a companion to cabbage, garlic, marjoram, rosemary, strawberries, and Swiss chard, but not fennel.
- **Leeks**: They don't always play nicely but will make friends with beets, carrots, celery, marjoram, mint and onions.
- **Lemon balm**: For best results, plant lemon balm with apples, apricots, cherry and other fruit trees like mulberry, roses, and zucchini.
- **Mint**: Mints are best with broccoli, brussels sprouts, cauliflower, cabbage, kale, mustard, radish, tomatoes, and zucchini. Keep mint away from chamomile and onions.
- **Mustard**: Mustard plays well with apples, apricots, cherry, fruit trees, grapevines, and tomatoes.
- **Parsnips**: A lovely vegetable that is also quite friendly. It is beneficial for beans, chives, coriander, corn, lettuce, marjoram, and onions. You might also like to plant peas, pennyroyals, raspberries, sage, and tomatoes. Avoid carrots and celery with parsnips.
- **Potatoes**: Plant potatoes with all of the beans except for climbing beans, and they will help broccoli, Brussels sprouts, cabbage, cauliflower, corn, marigolds, marjoram, peas, and tomatoes. They aren't good companions for apples, celery, cherry, cucumber, or radish.
- **Raspberry**: Plant marigolds, parsley, tansy and rue with raspberries but avoid planting with potatoes and blackberries.

- **Roses**: Despite their delicate look, roses are quite hardy. They get along with chives, marigolds, garlic, onions, parsley, peas, rue, and sage. The only thing you should do is keep roses away from tomato plants.
- **Shallots**: As a companion providing benefits to each other, really the best option is marjoram. Avoid shallots next to beans and peas.
- **Tansy**: This flower produces a lot of nectar and is a good companion for apples, apricots, borage, cabbage, cucumber, garlic, marigolds, and marjoram. It will benefit your roses, squash, and yarrow.
- **Thyme**: Apart from being extremely handy for the kitchen and a friend to many, thyme will accompany broccoli, collard greens, cauliflower. cabbage, kale, kohlrabi, mustard, radish and roses.
- **Tomato**: Plant tomatoes with asparagus, basil, beets, borage, carrots, cauliflower, celery, and chives. They will also pair well with grapevine, marigolds, marjoram, nasturtium, onions, parsley, and stinging nettles. Tomatoes don't play nicely with apricots, beets, fennel, potatoes, or rosemary.

My best advice here would be to make lists or a spreadsheet of the plants you want for your different zones and keep your maps handy. Also, remember that nothing is set in stone, and you don't have to plant everything straight away. You can start by choosing just a handful, and pay attention to feedback and then add more.

CROP ROTATION SYSTEMS

A well-cared-for fruit tree produces a greater yield each year, and the fruit becomes even more delicious each year. Nature doesn't always follow the same patterns, and it can be disheartening when the opposite occurs with our vegetables. Despite a great start, your green-fingered luck seems to have run out, but in reality, this isn't the case. Vegetables are fast growers and can use up certain nutrients more quickly than others. More often than not, the soil doesn't have enough time to recuperate the necessary nutrients for the next year. On the other hand, other nutrients in the same spot aren't being used, and this is the perfect reason to introduce crop rotation.

When we rotate crops, we take one plant, move it to a fresh spot and plant a new crop in the existing space. Ideally, you would have a minimum of three crops to rotate so that you never plant the same crop in a space for two years. There are two principal ways we can rotate our crops, either by family or by nutritional needs.

Crop rotation by family

When we rotate crops by family, we take advantage of the fact that all plants will have similar needs and traits. By keeping families together, you reduce the chance of disease spreading. So, if you have three garden beds, each with a different family, you would move the entire family into the next bed.

For example, in bed 1, you could plant beans and peas (the legume family), in bed 2 beets and onions (root crops), and in bed 3, cabbage and cauliflower (brassicas). In the second year, you would rotate your beans and peas into bed 2, beets and onions into bed 3, and your cabbage and cauliflower into bed 1. In the third year, you move each family into the garden bed they haven't been in yet. By the fourth year, your beans and peas can be planted back in their original garden bed.

For family crop rotation, don't plant the same family in a bed consecutively. You shouldn't plant another root crop in a bed that you have just used for your beets and onions. As you can imagine, coming from the same family, they will look for the same nutrients.

Crop rotation by nutritional needs

We can look at our crops as heavy feeders, heavy givers, and light feeders. It wouldn't make sense to plant two heavy feeders in a row as they will deplete the resources in the soil faster. After a heavy feeder, we plant a heavy giver like a nitrogen fixer to replenish the soil. For the third year, we would plant a light feeder that doesn't need a lot of nutrients, so the soil has a chance to rest before another year of heavy feeders. Here are some examples of feeders and givers:

- **Heavy feeders**: Generally speaking, these are our fast-growing crops, and because of their rapid growth, they need a lot of nutrients. For example, asparagus, beets, cabbage, lettuce, peppers, potatoes, tomatoes, strawberries, and watermelon.
- **Heavy givers**: Our nitrogen-fixers include your legumes, alfalfa, clover, beans, peas, peanuts, lentils, fenugreek, and soybeans.
- **Light feeders**: These are our root crops, such as carrots, garlic, leeks, mustard, onions, parsnips, and shallots.

With both family and nutritional rotation, you can use a container as your third garden bed if you are short of space. If you have a large space, you could have a fourth bed with a green manure crop that is cut down to get the nitrogen back into the soil or add compost and straw mulch. This fourth stage is another way to rest the soil for longer and give it an extra round of "giving."

Though this chapter is short, it prevents disaster and extra work in the long run by having to replant crops that haven't grown well, despite all of your efforts. It shows just how crucial planning every detail is. Imagine if you had only read about family crop rotation and decided to plant your tomatoes and potatoes together, then to discover that they aren't good companions. Hopefully, this chapter hasn't limited your options but instead has shown you just how many possibilities there are. Now that the designing stage is complete, it's necessary to learn how to maintain your permaculture garden and if the planning follows all of the principles and ethics, this should be nice and easy!

Chapter 12

Maintaining Your Permaculture Garden

I CAN'T RESIST ADDING MORE and more to my garden. The best bug you can catch is one that helps the environment. But you have to be careful. The more you add, the more maintenance there will be. If we think back to the first few chapters, the whole point is to create a design that takes care of itself, or at least as much as possible. Creating a self-automated garden is why many people call permaculture lazy gardening. The low maintenance makes it relatively easy, but considering the effort we put into the design, I wouldn't go as far as lazy!

While we have touched on most of the following points, it's easy to forget when we get carried away with our wildlife and plant companions. Here is a recap of eight tips that will make maintaining your permaculture garden a breeze and still achieve high yield.

Low-Maintenance Lazy Gardening Tips

No tilling

The life of your garden, the food for your food, we have to protect the soil. Tilling is like picking the scab off an unhealed wound. Leave the soil untilled so that you don't harm the microbes and upset the natural balance that is going on under that top layer.

Sheet mulching

Layering your mulch like lasagna helps to mimic the natural layers of the soil. Remember, the first layer should be overlapping cardboard or paper to prevent weeds popping up, then add compost or manure, topped with green mulch, kitchen scraps, and coffee grounds. The final layer is straw or wood chippings. There will be little need for weeding and you will have rich fertilized soil.

Plant your perennials

The only annuals you should consider planting are those crops that provide you with a resource (i.e. food). Annuals take work and time. Perennials just keep coming back, and many of them can also feed the family. Add some perennials like berries, kale, and artichoke, and we can't forget about the list of sixteen perennial plants for permaculture covered in Chapter 11.

Pot the sprouting pantry food

Instead of chucking things like potatoes, ginger, garlic, and onions onto your compost pile, regrow them. Depending on the time of year, you can plant them straight into the ground or into pots. You can do the same with some seeds such as chia, sunflower, and tomato seeds.

Make the most of your household scraps

If you aren't using your newspaper as a mulch layer, shred it and add it to the compost pile (remember, don't add glossy paper or cardboard). If you have a fish tank, reuse the water in your garden or to make compost tea as it will be rich in nitrogen. From banana peels to tea bags and crushed eggshells, all of your kitchen waste can be composted and made into nutrient-rich fertilizer.

Do you really need that lawn?

Lawns are great spaces for kids to play, but they require maintenance and lack biodiversity. Instead of being an unproductive space that needs mowing, could you convert it into raised beds or space for livestock?

Collect fallen leaves for mulch

The perfect example of nature doing its job; during fall, leaves wilt and cover the ground. During winter, they decompose, adding nutrients to your soil. Rake up leaves to add a thick layer of mulch and help protect the soil from the cold.

Collect every drop of water

Whether in barrels, swales, or rain gardens, make the most of all the rain that falls and remember to always plan for overflow. If you only have a small space, clay pots can save tons of work and water. Drip irrigation systems are cost-effective and easy to set up. Be sure to have easy access to water in all corners of your garden.

No natural ecosystem in the world requires humans to come in and carry out regular maintenance. Some people may think they are helping, but they are more likely messing up the natural balance. With this in mind, copy your ecosystem. It makes no sense to dig a pond in one area when water naturally collects in another. Look at a forest floor. Do you ever see bare soil? My permie mantra that should be sowed into your mind by now: bare soil is bad soil!

We have also mentioned in great detail the importance of biodiversity. Say no to monoculture designs in pretty little rows. The more you can plant, including taking advantage of vertical spaces, will create more interactions. The more interactions in a space, the more benefits each plant and animal will give to each other. Think sharing is caring. However, if you want to go low-maintenance, you can't just plant anything anywhere. Native crops, by design, will require less work. A little bit of research on what is native to your zone will save you a lot of time bent over in the garden.

All ecosystems have their weeds and wildlife. Placing cardboard down will help control the spread of weeds but those that are tough enough to make their way through deserve a chance to give their benefits to your garden. If you have a somewhat persistent weed, find a natural balance by adding an animal that sees the weed as delicious. Hopefully, Chapter 10 gave you plenty of inspiration for the insects and animals that can do a lot of the work for you. Chickens, rabbits, and sheep might be cute and add a bit of personality to your garden, but don't forget how beneficial a biodiverse range of insects can be.

The Best Low-Maintenance Crops

Whether you are an experienced permaculturist looking to add new elements or you are just starting to dip your spade in the soil, there is an excellent range of low-maintenance plants to choose from.

Most people like to plant annual crops like garlic, onions, potatoes, sweet potatoes, and winter squash because they can go into cold storage for approximately eight months. Crops like broccoli, Brussels sprouts, cabbage, and leeks require a little work after harvest, but you can take advantage of regular fresh vegetables because of their long season.

Choose your tomato plants wisely. Determinate tomatoes tend to flower once per season, and indeterminate tomatoes can produce fruit and grow throughout the season. Asparagus is a perennial, so it's high up on the low-maintenance list because once planted, it keeps giving for years.

Perennial herbs like chives, oregano, and thyme are easy to care for and add tons of nutrients and flavors to a wide range of dishes. The only care they need is a prune back to 5 inches of growth in late summer, early autumn. Although basil and cilantro are annuals, they are versatile, pretty much take care of themselves throughout the year, and are good pollinators.

If planted correctly and pruned once or twice a year, your fruit trees like apples, citrus fruits, and pears will provide you with years of crops. Berries are easy to grow, but they won't last long at room temperature. The best practice is to harvest your berries and freeze them straight away.

If you haven't got your garden calendar, it might be time to do it now. With everything going on in our lives, it is easy to forget a harvest window. Ideally, you will have planted different crops that will produce a yield all year round. Mark each harvest window on the calendar to know what's coming up, and you won't miss out. Plus, you'll know when you need to make a bit of extra space in your freezer. When planning this way, you can spend ten to fifteen minutes a day in your garden to keep an eye on everything and take in the feedback.

While there is not much new information in this chapter, it refreshes the permaculture principles. Sometimes, after reading up on plants and animals, we can get carried away, and before you know it, you have an ecosystem that is totally out of balance and takes an immense amount of hard work to maintain. Start with the water and soil, get this right, and a tremendous amount of work will be reduced in the long run. Then start to introduce crops and wildlife, all with a minimum of three functions and good company for each other.

Chapter 13

Inspiring a Permaculture Community

Never think that what you are doing in your garden won't make much of a difference. Edward Lorenz's butterfly effect showed that a tiny change in initial conditions had created a significantly different outcome. Be the butterfly! Start small in your home and then inspire one neighbor, then another, and before you know it, you have inspired a permaculture community.

Get Rid of the Waste

We spend so much time designing our garden because we want to create a system that mimics the natural world. Remember, nature doesn't produce waste, and because we pay so much attention to what nature tells us, permaculture systems should not produce a lot of waste, and ideally, there should be no waste produced. It's a shame that it will be a while before the rest of modern society catches up with the same values. Right now, the world is looking for ways to reuse waste, but isn't it more sensible not to create waste in the first place?

There are other ways that your permaculture garden is going to reduce waste and help the environment. Growing your food means fewer trips to the supermarket, your homegrown fruit and vegetables don't require plastic packaging, and you can recycle everyday household objects like toilet rolls, egg cartons, and pallets to make planters.

Before looking at how we can become less wasteful in our homes, here are a few ideas on sharing permaculture within the community. Any excess food should be shared with your friends and neighbors, or you can take your surplus to a local shelter or soup kitchen where it will be put to good use. You can also ask people in your community for their food waste. Just imagine how many food scraps you can collect from local restaurants or farmer's markets, all of which will turn into tons of nutrient-rich compost for your garden (another waste loop closed!).

Be sure to save the seeds from fruit and vegetables. You might be tempted to plant them all, but this will add to your maintenance. The idea is that you give away your surplus, not grow for the whole community (unless that's what you want to do and there is still no waste). Keep the seeds and share them with other permie enthusiasts in your area. Setting up a local seed exchange encourages biodiversity in all your gardens.

ALTERNATIVE WAYS TO SAVE WATER IN YOUR HOME

Hopefully, one of the first things you did, or are planning to do, is harvest rainwater. We also talked about how graywater can quench the thirst of many of our plants. Of course, we can't introduce human waste into our graywater, but flushing toilets with perfectly good drinking water is appalling! Approximately 30 percent of our household water gets flushed. Here is a quick look at how we can reduce the problem of human waste and save water.

- **Low water toilet flush**: Switching from a traditional toilet to a low water flush is enough to save a family of four around 80,000 gallons annually.
- **Solar-powered toilets**: Two permie birds with one stone, these solar-powered toilets can take human waste and turn it into hydrogen gas.
- **Biodigester**: This technology uses anaerobic bacteria to break down human and animal waste into biogas resources and turn waste into heat, fuel, and electricity.
- **Vermicomposting flush toilets**: Yes, earthworms will process human waste and toilet paper into compost.
- **Dry composting toilets**: Essentially, this is a bucket under a toilet seat with carbon-based materials like straw or cardboard. The waste is then added to a human waste compost pile.
- **Urine diversion toilet**: This is a toilet that separates urine and diverts it to be used as a natural fertilizer (8 parts water to 1 part urine). You can use urine to activate a compost pile or divert it to your graywater. Don't use urine from anyone who is taking antibiotics or other medications.

THE 7 RS OF A PERMACULTURE LIFESTYLE

Overall, we need to start being a little stricter with ourselves. We might want the latest smartphone but do we need it? Yes, it has some amazing functions, but so did your last phone, and did you use them all? An estimated 5.28 billion people have a mobile device. Imagine the waste we create every time we get a new one, and the same goes for laptops, TVs, and other electronics. Start saying no to waste and ask yourself, do I absolutely need a new device? Is it necessary to accept so much plastic from your supermarkets?

Once we get good at saying no to all that is not necessary, we can apply the 7 Rs:

- **Reduce**: Reduce all that we use from the faucet left open, the lights left on, the soap powder we use, the times we take the car when we could walk.
- **Repair**: Keep all warranties and repair things before disregarding them. If you know a handy person in your community, see if they can repair something in return for a crop you grow. If there seems like no other solution, watch some videos, and see if you can't tinker with it.
- **Repurpose**: One of the most original ideas that I have seen in a garden is someone who used an old fridge freezer as a raised garden bed—the perfect way to repurpose something that can't be repaired. In this case, a professional will have to remove the refrigerant first.
- **Reuse**: Thanks to many apps, secondhand goods no longer have the same negative reputation as before. Clothes, toys, furniture, and even things like old school textbooks can have a second or third home.
- **Rot**: Even if it's just a small compost bucket in your garden, you are making a difference. Try getting town halls and councils involved to create communal compost piles.
- **Recycle**: Anything that doesn't fall under the first 7 Rs can be recycled. Furthermore, ensure that the plastic and paper you buy come from recycled materials—closing another loop.

- **Rethink**: Is there any alternative to buying something. Can you borrow from others in the community rather than buy new, tempting them with a basket of your fresh goods. Can it be rented?

How You Can Be Part of the Solution

It's amazing how refreshing it is to see more and more people want to take on a more proactive approach to caring for the planet, reducing waste, and eating fresh, homegrown food. Like you before, they are possibly overwhelmed, unsure where to start, or worried about the time and effort it will take. Even though you don't feel like an expert just yet, you have one thing that they don't, and that is enthusiasm. Share your maps with others, talk about the local native, perennial or companion plants you want to introduce and how the chickens will keep your pesky insects under control.

If there aren't any local efforts to get more people involved, what can you do to get the ball rolling? Is there an empty patch of land that the city council has left bare? Can you drum up a load of signatures petitioning for a communal permaculture space? If you are an introvert and couldn't possibly imagine being the permaculture voice of your community, let your enthusiasm rub off on an extrovert who will get the ball rolling.

Fair share in your community

The third permaculture ethic ensures we all get a fair share, not growing or taking more than we need. So far, we have looked at this ethic from a harvest point of view—and collecting other scraps and leaves for compost. Here are just a few ideas that can bring a community together and create an even more spectacular permaculture community:

- Once a few of you are on a roll, rather than buying seeds, share seeds and cuttings. I absolutely love all the videos online that show you how to grow food from kitchen scraps. Set up a Facebook or WhatsApp group, and if you are looking for something new to grow, ask the group before you buy, "Anyone have some wine tomatoes? Can you save the seeds, please!" "Just cut down my sunflowers. Does anyone want some seeds?"

- Share your knowledge. Not everything is going to just miraculously grow. There will be some things that you grow very well and others that are going to need tweaking. Sharing your permaculture experiences is a great way to help others make improvements and for you to learn more before planting something.

- Share tools. Quality tools are essential for gardening, whether it's your spade or your wheelbarrow. Of course, with quality comes cost. But if you can share tools among you, you get to save money and get more use out of them.

- Permaculture babysitting. If your neighbors are going away, offer to garden sit for them and vice versa. The gardens will be well cared for, and you will also learn more from observing their ecosystem.

- Have a "permie party." I love permie parties because all the neighbors bring a dish made from their crops, giving us many inspiring ideas. We also get to share our ideas and experiences and, of course, socialize!

Not everyone will share your love for permaculture, and that's OK. But don't give up. Flap your wings hard enough, and you will make a difference. A lot of people still have the mindset that growing your food is more challenging than going to the store and picking up a plastic bag of chemically produced food. They haven't yet seen the bigger picture or the benefits of growing your food, such as increased health benefits, satisfaction, enhanced flavors, positive impact on the environment, and increased wellness from food. However, there are plenty of people who are ready and willing to learn.

AFTERWORD

Permaculture is nothing new. Way before man started to think he knew better, nature was taking pretty good care of itself. Each element provided the right amount of benefits to ensure survival and the survival for those who shared the environment. After humans began experimenting and making a mess of the natural balance, the philosophy of permaculture has risen again in an effort to repair some of the harm that has been done.

Yes, there is a lot to take in and a lot to consider in your plan, but everything will always come back to the twelve principles of permaculture:

- Observe and interact
- Catch and store energy
- Obtain a yield
- Apply self-regulation and accept feedback
- Use renewable resources
- Produce no waste
- Design from patterns to details
- Integrate rather than segregate

- Make slow, small changes
- Use and value diversity
- Use edges
- Use and respond to change

Combine your principles with the permaculture ethics of people care, fair share, earth care, and animal care, and you have the foundations for an amazing self-sustaining garden that mimics nature working at its best.

Another aspect that will help you maintain your garden with little effort while getting the best out of all elements is the rule of stacking functions. If something in your design doesn't have at least three functions, it needs to be rethought. With time, you will learn the inputs and outputs of each element but always research in those early days just to be on the safe side. The same can be said about your companion plants. The list is extensive, and getting it right the first time will save you time and energy in the future and give your permie confidence a good boost.

If you don't feed your garden, you can't expect your garden to feed you. There is absolutely no need for fertilizers when you have all you need right in your own home. Kitchen scraps, coffee grounds, newspapers, and plant cuttings after pruning can all be turned into nutritious food for your soil. If you don't have animals that can give you manure, visit a local riding school, they have plenty and often give it away—or trade it for some home-grown carrots and apples! Find every loop and make sure it's closed so that there is no waste from your home.

One of the biggest resources we waste is water. Just because it flows from a tap, it doesn't mean we can take it for granted. Take advantage of harvesting rainwater and ensure you have the right irrigation systems in place so no drop is wasted.

This book has numerous ideas for garden beds, and they can be created in a range of spaces, small and large. Bare soil is bad soil! Each space you have can be used for a crop, and polyculture is always the way to go. Whether looking at raised beds, plant guilds, or herb spirals, planting a diverse range of crops will prevent soil erosion, provide added nutrients, encourage pollinators, and keep pests away. All of this will make it easier for you to maintain your garden for years to come.

I feel like such a bore reminding you to be patient, but it is crucial. The planning stage will allow you to get the most out of each area you have in your garden. It will reduce the number of mistakes you make and ensure you use all of the available resources. Just because you aren't getting your hands in the dirt doesn't mean that observation and drawing your maps isn't taking you one step closer to being a fully-fledged permaculturist. Anyone with experience will tell you that the most successful permies take the time with the initial design.

One final reminder: Don't forget to check with the local city council about permits and get the plans for where your utility lines are. This is something that is often overlooked because of enthusiasm but could be a potential disaster if forgotten! If you can join a local permaculture community group, you may find someone who already has the answers to help you, and if not, you can share your resources and help others.

You now have all the information necessary to begin. It's been an absolute pleasure sharing what I have learned. Credit where credit is due, nature has been my best teacher, but I strive to keep learning by reaching out and learning from others every day. I would love to hear about all of your experiences, stories, and photos, and I invite everyone to join the Think Like an Ecosystem—Permaculture Guild Facebook Group so that like-minded people can continue to grow together.

Another way that we can help others to get excited about permaculture is to share reviews on Amazon. I would be forever grateful if you could leave a short review so that we can spread a little more enthusiasm and, bit by bit, create a greener, healthier planet.

Please use this link or scan the QR code to be sent directly to Amazon's review page.
http://www.amazon.com/review/create-review?&
asin=B09RGBCVW9

SCAN ME

A Free Gift For You!

FRUIT TREE PRUNING AND SHAPING GUIDE

5 SIMPLE STEPS TO PRUNE FRUIT TREES FOR HUGE HARVESTS AND EASY PICKING

AMÉLIE DES PLANTES

In the Fruit Tree Pruning and Shaping Guide, you'll learn...

- Why it is important to prune your fruit trees
- 5 Simple steps to prune your fruit trees for huge harvests

- How to choose the best shape for your fruit tree
- How to revive an abandoned or neglected fruit tree
- And so much more...

To receive your Fruit Tree Guide Scan the QR code or click the link

http://ecologicalfoodforest.com/fruit-tree-guide/

About the Author

AMÉLIE DES PLANTES

Amélie is a permaculture designer and environmentalist with interests in gardening, beekeeping and homeopathy. She has two children who thrive in the permaculture environment, and she loves watching them learn and grow alongside her. Amélie is passionate about expanding her knowledge and views her development as a permaculture designer as a lifelong journey.

Amélie is the founder of Ecological Food Forest, a site for permaculture education that welcomes members from across the world. She believes that bringing people together to share both triumphs and challenges can help make permaculture more comprehensive and accessible to the wider world. Amélie truly believes that permaculture practices have the power to change the world, and it is this conviction that pushes her to share her knowledge and passion with others.

References

Atkins, G. (2021, July 5). *The 12 Principles of Permaculture Explained*. The Homesteading Hippy. https://thehome-steadinghippy.com/principles-of-permaculture/

Bailey, P. (2019, May 28). *10 Benefits of Urban Permaculture*. Agritecture. https://www.agritecture.com/blog/2019/5/28/10-benefits-of-urban-permaculture

Bill Mollison. (2021b, July 22). Wikipedia. https://en.wikipedi-a.org/wiki/Bill_Mollison#Development_of_permaculture

Burrows, S. (2018, February 4). *Why Permaculture Is the Future of Food, If There Is a Future of Food*. Return to Now. https://re-turntonow.net/2017/12/11/permaculture-future-food/

Burtner, N. (2014, September 5). *How to Plan and Develop a Permaculture Site*. The Permaculture Research Institute. https://www.permaculturenews.org/2014/09/05/plan-develop-permaculture-site/

Butterfly Effect. (2021c, September 16). Wikipedia. https://en.wikipedia.org/wiki/Butterfly_effect

Carnegie Museum of Natural History. (2020, February 27). *Biomimicry Is Real World Inspiration*. https://carnegiemn-h.org/biomimicry-is-real-world-inspiration/

Cihon, B. (2020, September 8). *7 Easy Methods to Add Nitrogen to Your Soil*. Gardening Channel. https://www.gardeningchan-nel.com/add-nitrogen-garden-soil/

Community Fridges. (n.d.). Hubbub Foundation. https://www.hubbub.org.uk/the-community-fridge

Deep Green Permaculture. (n.d.). *Connecting People to Nature, Empowering People to Live Sustainably*. https://deepgreenper-maculture.com

Dyer, M. H. (2020, November 17). *The Sizes of Fruit Trees*. SFGATE. https://homeguides.sfgate.com/sizes-fruit-trees-57338.html

Eliades, A. (2016, June 3). *Perennial Plants and Permaculture*. The Permaculture Research Institute. https://www.permacul-turenews.org/2012/06/06/perennial-plants-and-permaculture/

Ellis, M. E. (n.d.). *Jasmine Training Guide—How to Train a Jasmine Vine*. Gardening Know How. https://www.gardening-knowhow.com/ornamental/flowers/jasmine/how-to-train-a-jasmine-vine.htm

Engels, J. (2015, May 14). *Why Permaculture Isn't Just Organic Farming*. One Green Planet. https://www.onegreenplanet.org/lifestyle/why-permaculture-isnt-just-organic-farming/

Enjoli, A. (2020, December 15). *What Happens If Bees Go Extinct?* LIVEKINDLY. https://www.livekindly.co/no-bees-no-food-its-that-simple/

Flores, H. J. (2019, April 5). *The 4 P's of Permaculture: Place, Patterns, Process, Principles* [Video]. YouTube. https://www.youtube.com/watch?v=YCzHIqljQ-Y

Flores, H. J. (2021, January 12). *Design Anything with the GOBRADIME Permaculture Design Process*. Medium. https://medium.com/permaculturewomen/thinking-outside-the-garden-box-using-the-gobradime-permaculture-design-process-f4b60c0025c7

Folk, E. (2020, October 26). *How to Turn Your Backyard into a Certified Wildlife Habitat*. The Permaculture Research Institute. https://www.permaculturenews.org/2020/11/04/how-to-turn-your-backyard-into-a-certified-wildlife-habitat/

Food Waste Is a Massive Problem—Here's Why. (2021, June 11). FoodPrint. https://foodprint.org/issues/the-problem-of-food-waste/

Foster, A. (2019, September 23). *7 Ways to Attract Pollinators*. Piedmont Environmental Alliance. https://www.peanc.org/7-ways-attract-pollinators

The Future of Permaculture, a Sustainable Way to Grow. (2018, September 5). Aid and International Development Forum (AIDF). http://www.aidforum.org/topics/food-security/the-future-of-permaculture-a-sustainable-way-to-grow/

Gibson, A. (2018, September 7). *How to Plant Out a Herb Garden*. The Micro Gardener. https://themicrogardener.com/how-to-plant-out-a-herb-garden/

Green Dream Project. (2019, June 19). *Better than Drip Irrigation? | Most Efficient Garden Irrigation System | Save up to 90% More Water* [Video]. YouTube. https://www.youtube.com/watch?v=Puaj_QPGVB8

Happen Films. (2018, June 27). *Living a Radically Simple Permaculture Life on 1/4 Acre | Creatures of Place* [Video]. YouTube. https://www.youtube.com/watch?v=rCRukvZE2Vk

Harness, J. (2019, November 18). *The List of Useful Insects*. Sciencing. https://sciencing.com/the-list-of-useful-insects-12341017.html

Harris, N. (2020, September 16). *6 Graphics Explain the Climate Feedback Loop Fueling US Fires*. World Resources Institute. https://www.wri.org/insights/6-graphics-explain-climate-feed-back-loop-fueling-us-fires

Hemenway, T. (2009) *Gaia's Garden: A Guide to Home-Scale Permaculture* (2nd ed.). Chelsea Green Publishing.

Horvath, W. (2015, October 9). *The Definitive Guide to Building Deep Rich Soils by Imitating Nature*. Permaculture Apprentice. https://permacultureapprentice.com/building-soil/

Horvath, W. (2017, March 17). *How to Read the Landscape*. Permaculture Apprentice. https://permacultureapprentice.-com/how-to-read-the-landscape/

How & Why to Make Actively Aerated Compost Tea to Feed Your Garden. (2021, April 10). Homestead and Chill. https://home-steadandchill.com/actively-aerated-compost-tea/

How to Survey a Site. (2012, July 21). Learn Permaculture. http://www.learnpermaculture.com/blog/74-permaculture-design-pm-2

In the Garden. (2020, July 4). *Installing Drip Irrigation in Vegetable Garden | A Beginners Guide to Drip Irrigation* [Video]. YouTube. https://www.youtube.com/watch?v=MQNdGFI9wiQ

Jenkins, E. (2020, October 5). *7 Ways to Harvest Rainwater for Your (Permaculture) Garden*. New Life on a Homestead. https://www.newlifeonahomestead.com/harvest-rainwater-for-your-permaculture-garden/

Jenkins, E. (2021, August 4). *Planting Permaculture Guilds— Your Comprehensive Guide*. New Life on a Homestead. https://www.newlifeonahomestead.com/permaculture-guilds/

KCET. (2017, February 9). *Tending the Wild: Complete Broadcast Special* [Video]. YouTube. https://www.youtube.com/watch?v=TbxLv9EEzs8

Köchel, P. (2019, July 15). *What If All Houses Were Covered with Solar Panels?* INSH. https://insh.world/science/what-if-all-houses-were-covered-with-solar-panels/

List of Beneficial Weeds. (2021a, July 12). Wikipedia. https://en.wikipedia.org/wiki/List_of_beneficial_weeds

Lovely Greens. (2020, June 10). *DIY Herb Spiral—Clever Way to Grow Lots of Herbs in a Small Space* [Video]. YouTube. https://www.youtube.com/watch?v=T4RtAzb88R8

Madrigal, A. (2008, December 29). *From Salon to Salad: Human Hair Makes Good Plant Fertilizer*. Wired. https://www.wired.com/2008/12/hairmats/

Manchee, L., & Manchee, K. (n.d.). *Integrating Animals— Permaculture*. Keela Yoga Farm. https://www.keelayogafarm.com/permaculture/integrating-animals-on-a-permaculture-farm/

Marx, M. (2019, March 28). *Permaculture Principle 1—Observe & Interact*. Gympie & District Landcare. https://gympielandcare.org.au/permaculture-principle-1-observe-interact/

Merklein, M., & Colorado, M. (n.d.). *The Waste Cycle*. Perma-culture Women. https://www.permaculturewomen.com/zero-waste-permaculture-class/

Mollison, B. C. (1988). *Permaculture: A Designers' Manual*. Tagari Publications.

Mollison, B. (2013). *Introduction to Permaculture* (2nd ed.). Tagari Publications.

Nautiyal, C. S., Chauhan, P. S., & Nene, Y. L. (2007, December 3). Medicinal Smoke Reduces Airborne Bacteria. *Journal of Ethnopharmacology,* *114*(3), 446–451. https://doi.org/10.1016/j.jep.2007.08.038

Noonan, J. (2016, August 19). *10 Ways Your Backyard Can Hurt You*. Bob Vila. https://www.bobvila.com/slideshow/10-ways-your-backyard-can-hurt-you-50383

Open Source Ecology. (2011, March 12). *Inputs and Outputs in Permaculture—Open Source Ecology*. https://wiki.opensource-ecology.org/wiki/Inputs_and_outputs_in_permaculture

General Mills (2021, April). *Global Responsibility 2021*. https:/globalresponsibility.generalmills.com/HTML1/default.htm

Outdoor Water Use in the United States. (n.d.). WaterSense—United States Environmental Protection Agency. https://19jan-uary2017snapshot.epa.gov/www3/watersense/pubs/outdoor.html

Permaculture Guilds. (2000). Never Ending Food. http://www.neverendingfood.org/b-what-is-permaculture/per-maculture-guilds/

Plant Abundance. (2014, September 28). *DIY Backyard Rain-water Harvesting Using Repurposed Food Grade Barrels* [Video]. YouTube. https://www.youtube.com/watch?v=ZC7A3621_hg

Poore, J., & Nemecek, T. (2018, June 1). Reducing Food's Environmental Impacts through Producers and Consumers. *Science,* *360*(6392), 987–992. https://www.science.org/doi/10.1126/science.aaq0216

Powers, M. (2017). *The Permaculture Student 2: A Collection of Regenerative Solutions* (2nd ed.). PermaculturePowers123.

Prescott, L. (2019, January 15). *Permaculture: A Whole Design Philosophy for Sustainable Living.* Medium. https://medium.com/colab-dudley/permaculture-a-whole-design-philosophy-for-sustainable-living-7204f0a57edb

Raised Beds. (2013, April 22). Ecologia Design. https://www.ecologiadesign.com/2013/04/22/raised-beds/

Ritchie, H. (2019, November 6). *Food Production Is Responsible for One-Quarter of the World's Greenhouse Gas Emissions.* Our World in Data. https://ourworldindata.org/food-ghg-emissions

Roberts, T. (2017, October 12). *A Primer on Creating Soil.* The Permaculture Research Institute. https://www.permaculturenews.org/2017/10/13/primer-creating-soil/

Sampson-Kelly, A. (2016, February 13). *Biomimicry & Permaculture Today.* Permaculture Visions. https://permaculturevisions.com/biomimicry-permaculture-today/

SanSone, A. (2020, April 27). *11 Plant Combos You Should Grow Side-by-Side.* Country Living. https://www.countryliving.com/gardening/news/g4188/companion-planting/

Sayner, A. (2021, July 1). *16 Permaculture Plants You Should Have in Your Garden.* GroCycle. https://grocycle.com/permaculture-plants/

Schauder, N. (2020, March 24). *A Medicinal Herb Garden*. Permaculture Gardens. https://growmyownfood.com/medicinal-herb-garden/

Sirbu, E. R., Margen, S., & Galloway, D. H. (1967, November 1*)*. Effect of Reduced Protein Intake on Nitrogen Loss from the Human Integument. *The American Journal of Clinical Nutrition,* *20*(11), 1158–1165. https://doi.org/10.1093/ajcn/20.11.1158

Sorensen, J. (2020, October 10). *The Complete Beginner's Guide to Greywater Systems*. Elemental Green. https://elemental.green/complete-beginner-guide-to-greywater-systems/

Stacking Functions in Permaculture. (2016, April 8). A Floresta Nova. https://aflorestanova.wordpress.com/2016/04/08/stacking-functions-in-permaculture/

Starhawk. (2016, December 1). *Social Permaculture—What Is It?* Foundation for Intentional Community. https://www.ic.org/social-permaculture-what-is-it/

Stross, A. (2021a, August 23). *How to Build a Swale in the Residential Landscape [+ Free Download]*. Tenth Acre Farm. https://www.tenthacrefarm.com/how-to-build-swale/

Stross, A. (2021b, September 8). *Here's a Quick Way to Terrace a Hill [+ Free Download]*. Tenth Acre Farm. https://www.tenthacrefarm.com/quick-terrace-hill/

Tammy P. (2020, September 12). *How to Boost Nitrogen in Soil with Hair*. Do It Yourself. https://www.doityourself.com/stry/how-to-boost-nitrogen-in-soil-with-hair

Tanner, M. (2016, May 10). *A Better Shade of Grey: Grey Water in Your Garden*. Renew. https://renew.org.au/sanctuary-magazine/ideas-advice/a-better-shade-of-grey-grey-water-in-your-garden/

Thornbro, H. (2021, August 14). *How to Use Cover Crops in Your Permaculture Garden*. Redemption Permaculture. https://redemptionpermaculture.com/how-to-use-cover-crops-in-your-permaculture-garden/

The 12 Principles of Permaculture. (2019, February 6). live-native.com. https://www.live-native.com/the-12-principles-of-permaculture/

University of California, Division of Agriculture and Natural Resources. (n.d.). *Disease Focus: Weeds as a Source of Plant Virus Infections and Bacterial Leaf Spot of Poinsettia*. http://ucnfanews.ucanr.edu/Articles/Disease_Focus/DISEASE_FOCUS__Weeds_as_a_source_of_plant_virus_infections_and_bacterial_leaf_spot_of_poinsettia/

Waddington, E. (2020, May 29). *40 Nitrogen Fixing Plants to Grow in Your Garden*. Rural Sprout. https://www.ruralsprout.com/nitrogen-fixing-plants/

Washburn, J. (2021, June 9). *Living Off the land: How to Start a Permaculture Garden*. [your]NEWS Media Group, Inc.

https://yournews.com/2021/06/09/2149354/living-off-the-land-how-to-start-a-permaculture-garden/

Westmoreland, P., & Halsey, D. (2001). *Natural Capital Plant Database*. Natural Capital. https://permacultureplantdata.com

What Is Permaculture? Principles, Benefits, and Facts. (2021, August 18). PermaOrganic. https://permaorganic.com/what-is-permaculture-principles-benefits-and-facts/#Benefits_of_Permaculture

What Is Regenerative Agriculture? (2019, July 2). Climate Reality. https://www.climaterealityproject.org/blog/what-regenerative-agriculture

Williams, A. (2020, September 30). *How to Build a French Drain*. WikiHow. https://www.wikihow.com/Build-a-French-Drain

Wooldridge, T. (2016, April 13). *Clay Pot Irrigation*. Permaculture Food Forest. https://permaculturefoodforest.wordpress.com/2016/04/13/clay-pot-irrigation/